# Methods in Molecular Biology

*Series Editor*
**John M. Walker**
**School of Life and Medical Sciences**
**University of Hertfordshire**
**Hatfield, Hertfordshire, UK**

For further volumes:
http://www.springer.com/series/7651

For over 35 years, biological scientists have come to rely on the research protocols and methodologies in the critically acclaimed *Methods in Molecular Biology* series. The series was the first to introduce the step-by-step protocols approach that has become the standard in all biomedical protocol publishing. Each protocol is provided in readily-reproducible step-by step fashion, opening with an introductory overview, a list of the materials and reagents needed to complete the experiment, and followed by a detailed procedure that is supported with a helpful notes section offering tips and tricks of the trade as well as troubleshooting advice. These hallmark features were introduced by series editor Dr. John Walker and constitute the key ingredient in each and every volume of the *Methods in Molecular Biology* series. Tested and trusted, comprehensive and reliable, all protocols from the series are indexed in PubMed.

# Diabetic Retinopathy

## Methods and Protocols

Edited by

**Guei-Sheung Liu and Jiang-Hui Wang**

*Centre for Eye Research Australia, University of Melbourne, East Melbourne, VIC, Australia*

Humana Press

*Editors*
Guei-Sheung Liu
Centre for Eye Research Australia
University of Melbourne
East Melbourne, VIC, Australia

Jiang-Hui Wang
Centre for Eye Research Australia
University of Melbourne
East Melbourne, VIC, Australia

ISSN 1064-3745 ISSN 1940-6029 (electronic)
Methods in Molecular Biology
ISBN 978-1-0716-3254-3 ISBN 978-1-0716-3255-0 (eBook)
https://doi.org/10.1007/978-1-0716-3255-0



This Humana imprint is published by the registered company Springer Science+Business Media, LLC, part of Springer Nature.
The registered company address is: 1 New York Plaza, New York, NY 10004, U.S.A.

# Preface

Diabetic retinopathy is a chronic, progressive disease and a leading cause of vision impairment and blindness. It is well established that diabetic retinopathy is strongly associated with the duration and extent of hyperglycemia, yet several underlying mechanisms of the disease remain unclear. Thus, early diagnosis and treatment options for this condition are currently limited. The aim of this *Methods in Molecular Biology* volume is to provide various cellular and experimental animal models, as well as research methodologies to understand the complex interrelated pathophysiological mechanisms triggered by hyperglycemia underlie the development of diabetic retinopathy. The potential of innovative technologies to uncover potential genetic factors or biomarkers is also highlighted here to advance the development of diabetic retinopathy management.

This volume comprises chapters that cover the research methods in detail for understanding the pathogenesis of diabetic retinopathy, detection and screening of biomarkers in diabetic retinopathy, as well as novel technologies for diabetic retinopathy management.

We would like to thank all the authors of this volume who have shared their expertise and protocols. We also wish to thank Dr. John M. Walker for his support during the process of completing this book. We hope this volume will be an invaluable resource for scientists and clinicians who are interested in the area of ophthalmology research.

***East Melbourne, VIC, Australia***

***Guei-Sheung Liu***
***Jiang-Hui Wang***

# Contents

## Contributors

BRIANNA C. AFIAT • *Department of Optometry and Vision Sciences, University of Melbourne, Parkville, VIC, Australia*

EMILY ATTRILL • *Tasmanian School of Medicine, University of Tasmania, Hobart, TAS, Australia*

VELUCHAMY AMUTHA BARATHI • *Translational Pre-Clinical Model Platform, Singapore Eye Research Institute, Singapore, Singapore; ACP in Ophthalmology & Visual Sciences, DUKE-NUS Graduate Medical School, Singapore, Singapore; Department of Ophthalmology, Yong Loo Lin School of Medicine, National University of Singapore, Singapore, Singapore*

BANG V. BUI • *Department of Optometry and Vision Sciences, University of Melbourne, Parkville, VIC, Australia*

KATHRYN P. BURDON • *Menzies Institute for Medical Research, University of Tasmania, Hobart, TAS, Australia*

NICK CARRUTHERS • *Bioinformatics Core, Biomedical Research Core Facilities, University of Michigan Medical School, Ann Arbor, MI, USA*

JIANSU CHEN • *Aier School of Ophthalmology, Central South University, Changsha, Hunan Province, People's Republic of China; Aier Eye Institute, Changsha, Hunan Province, People's Republic of China; Changsha Aier Eye Hospital, Changsha, Hunan Province, People's Republic of China*

SARANTIS CHLAMYDAS • *Olink Proteomics AB, Uppsala, Sweden*

YU-FAN CHUANG • *School of Respiratory Therapy, College of Medicine, Taipei Medical University, Taipei, Taiwan; Department of Pharmacology, School of Medicine, College of Medicine, Taipei Medical University, Taipei, Taiwan*

ZEKAI CUI • *Aier Eye Institute, Changsha, Hunan Province, People's Republic of China*

SRUJANIKA DASH • *Department of Bioinformatics, Odisha University of Agriculture and Technology, Bhubaneswar, Odisha, India*

CHENGCHENG DING • *Aier Eye Institute, Changsha, Hunan Province, People's Republic of China*

RYAN J. FARR • *Diabetes and Islet Biology Group, School of Medicine, Western Sydney University, Campbelltown, NSW, Australia; Health and Biosecurity (H&B) at the Australian Centre for Disease Preparedness (ACDP), CSIRO, Geelong, VIC, Australia*

CHRIS GATES • *Bioinformatics Core, Biomedical Research Core Facilities, University of Michigan Medical School, Ann Arbor, MI, USA*

JIANING GU • *Aier Eye Institute, Changsha, Hunan Province, People's Republic of China*

RAJYA L. GURUNG • *Menzies Institute for Medical Research, University of Tasmania, Hobart, TAS, Australia*

ANANDWARDHAN A. HARDIKAR • *Diabetes and Islet Biology Group, School of Medicine, Western Sydney University, Campbelltown, NSW, Australia*

HRISHIKESH P. HARDIKAR • *Diabetes and Islet Biology Group, School of Medicine, Western Sydney University, Campbelltown, NSW, Australia*

ANH HOANG • *Department of Optometry and Vision Sciences, University of Melbourne, Parkville, VIC, Australia*

YINHUA HUANG • *Aier School of Ophthalmology, Central South University, Changsha, Hunan Province, People's Republic of China; Aier Eye Institute, Changsha, Hunan Province, People's Republic of China*

SHANGLI JI • *Aier Eye Institute, Changsha, Hunan Province, People's Republic of China*

MUGDHA V. JOGLEKAR • *Diabetes and Islet Biology Group, School of Medicine, Western Sydney University, Campbelltown, NSW, Australia*

SATHEESH KUMAR • *Centre for Eye Research Australia, Royal Victorian Eye and Ear Hospital, East Melbourne, VIC, Australia; Ophthalmology, Department of Surgery, University of Melbourne, East Melbourne, VIC, Australia; Menzies Institute for Medical Research, University of Tasmania, Hobart, TAS, Australia*

POOJA S. KUNTE • *Diabetes and Islet Biology Group, School of Medicine, Western Sydney University, Campbelltown, NSW, Australia; Diabetes Unit, KEM Hospital and Research Centre, Pune, India*

RICARDO LAMY • *Department of Ophthalmology, University of California, San Francisco, San Francisco, CA, USA; Department of Ophthalmology, Zuckerberg San Francisco General Hospital, San Francisco, CA, USA*

PEI YING LEE • *Department of Optometry and Vision Sciences, University of Melbourne, Parkville, VIC, Australia*

YUJIN LEE • *Department of Ophthalmology, Inje University Busan Paik Hospital, Busan, South Korea*

HETIAN LEI • *Shenzhen Eye Hospital, Shenzhen Eye Institute, Jinan University, Shenzhen, China; Department of Ophthalmology, The Third Affiliated Hospital of Xinxiang Medical University, Xinxiang, China*

FAN-LI LIN • *Department of Pharmacology, School of Medicine, College of Medicine, Taipei Medical University, Taipei, Taiwan; School of Post-Baccalaureate Medicine, College of Medicine, Kaohsiung Medical University, Kaohsiung, Taiwan; Department of Pharmacology, College of Medicine, Kaohsiung Medical University, Kaohsiung, Taiwan*

GUEI-SHEUNG LIU • *Centre for Eye Research Australia, Royal Victorian Eye and Ear Hospital, East Melbourne, VIC, Australia; Menzies Institute for Medical Research, University of Tasmania, Hobart, TAS, Australia; Ophthalmology, Department of Surgery, University of Melbourne, East Melbourne, VIC, Australia*

JACEY HONGJIE MA • *Aier Eye Institute, Changsha, Hunan Province, People's Republic of China*

SHOWGY MA'AYEH • *Olink Proteomics AB, Uppsala, Sweden*

BENNET J. MCCOMISH • *Menzies Institute for Medical Research, University of Tasmania, Hobart, TAS, Australia*

SRILATHA MUNDLA • *Department of Biotechnology, Sri Venkateswara University, Tirupati, AP, India*

CHRISTINE T. O. NGUYEN • *Department of Optometry and Vision Sciences, University of Melbourne, Parkville, VIC, Australia*

TOMOHARU NISHIMURA • *Department of Ophthalmology, Dokkyo Medical University Saitama Medical Center, Saitama, Japan*

NAINA PATYAL • *Department of Bioinformatics, Central University of Himachal Pradesh, Kangra, HP, India*

YANFANG PENG • *Aier Eye Institute, Changsha, Hunan Province, People's Republic of China*

Mya Polkamp • *Diabetes and Islet Biology Group, School of Medicine, Western Sydney University, Campbelltown, NSW, Australia*
Dino Premilovac • *Tasmanian School of Medicine, University of Tasmania, Hobart, TAS, Australia*
Stephen M. Richards • *Tasmanian School of Medicine, University of Tasmania, Hobart, TAS, Australia*
Renee M. Ross • *Tasmanian School of Medicine, University of Tasmania, Hobart, TAS, Australia*
Madhu Sudhana Saddala • *Wilmer Eye Institute, Johns Hopkins University School of Medicine, Baltimore, MD, USA*
Jay M. Stewart • *Department of Ophthalmology, University of California, San Francisco, San Francisco, CA, USA; Department of Ophthalmology, Zuckerberg San Francisco General Hospital, San Francisco, CA, USA*
Jeffrey Sundstrom • *Department of Ophthalmology, Penn State College of Medicine, Hershey, PA, USA*
Brad A. Sutherland • *Tasmanian School of Medicine, University of Tasmania, Hobart, TAS, Australia*
Shibo Tang • *Aier School of Ophthalmology, Central South University, Changsha, Hunan Province, People's Republic of China; Aier Eye Institute, Changsha, Hunan Province, People's Republic of China; Changsha Aier Eye Hospital, Changsha, Hunan Province, People's Republic of China*
Ru Qi Teh • *Centre for Eye Research Australia, Royal Victorian Eye and Ear Hospital, East Melbourne, VIC, Australia*
Katie K. N. Tran • *Department of Optometry and Vision Sciences, University of Melbourne, Parkville, VIC, Australia*
Sai Bo Bo Tun • *Translational Pre-Clinical Model Platform, Singapore Eye Research Institute, Singapore, Singapore; Karolinska Institutet, Stockholm, Sweden*
Anna K. van Koeverden • *Department of Optometry and Vision Sciences, University of Melbourne, Parkville, VIC, Australia*
Jiang-Hui Wang • *Centre for Eye Research Australia, Royal Victorian Eye and Ear Hospital, East Melbourne, VIC, Australia; Ophthalmology, Department of Surgery, University of Melbourne, East Melbourne, VIC, Australia*
Yini Wang • *Aier Eye Institute, Changsha, Hunan Province, People's Republic of China; Aier School of Ophthalmology, Central South University, Changsha, Hunan Province, People's Republic of China*
Zhijie Wang • *Aier Eye Institute, Changsha, Hunan Province, People's Republic of China; School of Ophthalmology and Optometry, Eye Hospital, Wenzhou Medical University, Wenzhou, People's Republic of China*
Sarah Weber • *Department of Ophthalmology, Penn State College of Medicine, Hershey, PA, USA*
Vickie H. Y. Wong • *Department of Optometry and Vision Sciences, University of Melbourne, Parkville, VIC, Australia*
Wilson K. M. Wong • *Diabetes and Islet Biology Group, School of Medicine, Western Sydney University, Campbelltown, NSW, Australia*

WENYI WU • *Department of Ophthalmology, Xiangya Hospital, Central South University, Changsha, China; Department of Ophthalmology, Hunan Key Laboratory of Ophthalmology, Changsha, China; National Clinical Research Center for Geriatric Disorders, Xiangya Hospital, Central South University, Changsha, China*

CHITTARANJAN S. YAJNIK • *Diabetes Unit, KEM Hospital and Research Centre, Pune, India*

JAEWOOK YANG • *Department of Ophthalmology, Inje University Busan Paik Hospital, Busan, South Korea*

XU YANG • *Aier Eye Institute, Changsha, Hunan Province, People's Republic of China*

YONG ZENG • *Aier Eye Institute, Changsha, Hunan Province, People's Republic of China; Department of Ophthalmology, Sichuan Provincial People's Hospital, University of Electronic Science and Technology of China, Chengdu, People's Republic of China*

DA ZHAO • *Department of Optometry and Vision Sciences, University of Melbourne, Parkville, VIC, Australia*

YUANJUN ZHAO • *Department of Ophthalmology, Penn State College of Medicine, Hershey, PA, USA*

# Chapter 1

# Induction of Type 2 Diabetes in Mice to Understand Vascular Changes That Drive Diabetic Retinopathy

**Emily Attrill, Stephen M. Richards, Renee M. Ross, Brad A. Sutherland, and Dino Premilovac**

## Abstract

Diabetic retinopathy is a common complication of type 2 diabetes. Research into this comorbidity is challenging due to the slow progression of pathological changes and the limited transgenic models available to study disease progression and mechanistic changes. Here, we describe a non-transgenic mouse model of accelerated type 2 diabetes using a high-fat diet in combination with streptozotocin delivered via osmotic mini pump. This model, when subjected to fluorescent gelatin vascular casting, can be used to study vascular changes in type 2 diabetic retinopathy.

**Key words** Type 2 diabetes, High-fat diet, Streptozotocin, Insulin resistance, Retinopathy, Obesity, Vascular cast

## 1 Introduction

Type 2 diabetes (T2D) is a progressive metabolic disorder affecting approximately 380 million people globally [1]. It involves the slow (10+ years) transition from peripheral insulin resistance to T2D and the subsequent development of various comorbidities including cardiovascular disease, retinopathy, nephropathy, and neuropathy. Retinopathy is the most common diabetic comorbidity, currently affecting around 100 million diabetic patients globally [2]. Changes in microvascular structure and function drive the development of diabetic retinopathy; however, ascertaining the exact contribution of vascular pathology to disease is difficult primarily due to the long timeframe over which such changes occur. Consequently, we have developed a rodent model of accelerated T2D to assess microvascular changes in diabetic retinopathy.

The accelerated model of T2D progression utilizes both a high-fat diet and the DNA-alkylating agent streptozotocin (STZ) [3–5] to bring together three salient features of human T2D: insulin

Guei-Sheung Liu and Jiang-Hui Wang (eds.), *Diabetic Retinopathy: Methods and Protocols*, Methods in Molecular Biology, vol. 2678, https://doi.org/10.1007/978-1-0716-3255-0_1,

resistance, insulin deficiency, and moderate hyperglycemia. Once in the blood, STZ enters cells via the GLUT2 transporter [6], which is abundant on the surface of insulin-producing pancreatic beta cells. This makes the beta cells susceptible to STZ-induced cell death as a result of DNA alkylation, reducing insulin production by the pancreas. In Sprague Dawley rats, this combination of high-fat diet and STZ results in sustained but moderate hyperglycemia [4], which is the leading risk factor for the development of diabetic comorbidities [7–9]. Importantly, we have demonstrated the development of retinopathy and peripheral neurodegeneration in this rat model in the context of T2D [3–5].

Given that the combination of high-fat diet and STZ effectively accelerates the development of T2D in rats, we have subsequently adapted this model for use in mice to expand the number of transgenic options and molecular techniques to further investigate disease progression. Additionally, we have combined this novel T2D mouse model with a technique to generate a fluorescent vascular cast to investigate pathophysiological changes to the vasculature in diabetic retinopathy. Thus, the following protocol outlines how to efficiently establish moderate T2D in mice that, in combination with fluorescent vascular casting, enables the detailed interrogation of microvascular changes in the retina (and other organs) in the context of T2D.

## 2 Materials

Prepare all solutions/equipment using aseptic technique. All surgical equipment will need to be autoclaved and handled appropriately to maintain aseptic technique. Prepare all reagents at room temperature using sterile saline (0.9% NaCl) unless otherwise detailed. The dose of STZ required to induce T2D (rather than complete insulin deficiency or T1D) will vary slightly between mouse strains and will need to be confirmed by individual researchers. Prior to starting any animal work, ensure all procedures on animals are approved by an Institutional Animal Ethics Committee.

### 2.1 General Animal Husbandry and Monitoring Requirements

1. Adult, male C57BL/6 mice at 12–15 weeks of age.
2. Standard rodent chow: The fat content of this diet is 5.6% by weight.
3. High-fat diet (HFD): The fat content of this diet is 23% by weight, and complex carbohydrates have been replaced with sucrose.
4. Laboratory scales for weighing mice.
5. Glucose monitor and glucose test strips for routine, non-fasting blood glucose assessment.
6. Surgical-grade scissors for tail tip cut.

### 2.2 Osmotic Mini Pump Preparation, Implant, and Removal

1. Two sterile plastic sheets to create a sterile field.
2. Osmotic mini pump (Model 1002): This model has a 100 μL reservoir, a regulator that delivers 0.25 μL/h, and a blunt 27G needle to load the reservoir.
3. Citrate-buffered saline (0.1 mmol/L, pH 4.4). For solution A, dissolve 1.05 g citric acid monohydrate in 50 mL of sterile saline. For solution B, dissolve 1.47 g tri-sodium citrate dihydrate in 50 mL of sterile saline. Solutions A and B can be stored at 4 °C for up to 2 weeks. To make citrate-buffered saline at pH 4.4, mix solutions A and B in a 1:1 ratio to the desired volume.
4. Streptozotocin (STZ): Total dose required is 200 mg/kg delivered over 14 days (*see* **Note 1**), for example, for a 30 g mouse = 200 mg/kg × 0.03 kg = 6 mg in 100 μL citrate-buffered saline. Because the reservoir of the osmotic mini pump is 100 μL, make a total of 150 μL of the 200 mg/kg solution (9 mg) to ensure complete filling of the reservoir.
5. 15 mL falcon tube with approximately 3 mL sterile saline. This is used to store the loaded osmotic mini pump prior to implant (*see* **Note 2**).
6. Isoflurane vaporizer with induction chamber and mouse nose.
7. Long-acting analgesic: Meloxicam (5 mg/mL) injected subcutaneously at a dose of 5 mg/kg. For example, for a 30 g mouse, 30 μL of the 5 mg/mL solution is required to achieve the correct analgesic dose.
8. Rodent shaver to remove fur at the surgical site.
9. Povidone-iodine solution (10% w/v) for cleaning surgical sites.
10. Sterile surgical tools: Two pairs of rat tooth forceps, surgical scissors, and a small hemostat.
11. Michel wound clips: 9 mm wound clips with applier and remover.

### 2.3 Glucose Tolerance Testing

1. 30% glucose solution: Add 3 g D-glucose to 7 mL sterile saline. Dissolve and then adjust volume with additional saline to a final volume of 10 mL (*see* **Notes 3** and **4**). Use at room temperature.
2. 1 mL insulin syringes with a 27G needle.
3. Sterile surgical scissors.
4. Topical analgesic cream: 5% lidocaine.
5. Glucose monitor and glucose test strips. The exact number of glucose strips required will depend on the frequency of sampling.

#### *2.4 Cardiac Perfusion with Fluorescent Gelatin Vascular Casting*

1. Peristaltic pump with tubing and blunted 23G needle. The setup needs to achieve a flow rate of 7.5–8.0 mL/min for appropriate intravascular perfusion.
2. Angled stage with anchorages for all four limbs.
3. Surgical tools: Forceps, scissors, and hemostat (*see* **Note 5**).
4. Terminal anesthesia: Sodium pentobarbitone at a concentration of 65 mg/mL. One animal at a time is injected with a lethal dose of pentobarbitone (≥180 mg/kg). For a 30 g mouse, this equals 180 mg/kg × 0.03 kg = 5.4 mg. To obtain this dose from the stock solution, 5.4 mg ÷ 65 mg/mL = 0.083 mL = 83 μL.
5. 1× heparinized phosphate-buffered saline (PBS): First, make a 10× PBS stock solution by dissolving 80 g NaCl, 2.0 g KCl, 14.4 g $Na_2HPO_4$, and 2.4 g $KH_2PO_4$ in 800 mL of deionized water. Adjust pH to 7.4 using HCl and NaOH as required. Top up with additional deionized water for a final volume of 1 L. To make a 1× PBS solution, add 100 mL of 10× PBS to 900 mL of deionized water. Finally, add 1 mL of 5000 IU heparin to 1 L of 1× PBS. Store in a 4 °C fridge until required. Use at room temperature.
6. 4% paraformaldehyde (PFA) solution: Heat 800 mL of 1× PBS to 60 °C using a hotplate with a magnetic stirrer. Once the solution reaches 60 °C, add 40 g paraformaldehyde and dissolve. Add ~10 mL of 10 M NaOH until the solution goes from milky white to clear. Vacuum-filter the solution using 0.22 μm filter. Add 10 M HCl until pH 7.1. Adjust the final volume to 1 L using 1× PBS. Store in a 4 °C fridge until required. Use at room temperature.
7. Fluorescent gelatin for vascular casting. PBS solution with 1.25% bovine gelatin and 0.5% fluorescein isothiocyanate-dextran (molecular weight: 70,000): Heat 50 mL 1× PBS to approximately 60 °C. Add 625 mg bovine gelatin and mix until dissolved. Weigh 250 mg of FITC-dextran, and add to the solution immediately before cardiac perfusion. Keep the solution in a 37 °C water bath to maintain optimal temperature for perfusion and prevent gelatin setting (*see* **Note 6**).

## 3 Methods

#### *3.1 General Animal Housing and Monitoring*

1. Prior to the experiment, maintain mice on standard laboratory chow.
2. At week 0, make an initial recording of body weight and non-fasting glucose. To assess blood glucose, apply a small amount of anesthetic cream to the tail, covering ~1 cm back

from the tip. Leaving the mouse in the home cage for ~5 min, make a small cut to remove the tip of the tail (<1 mm, enough for a ~ 10 μL blood sample). Discard the first small drop of blood. Gently squeeze the tail and apply the second drop of blood to the test strip and record blood glucose concentration. Once complete, replace standard chow with the HFD and provide ad libitum. This marks the start of the experiment.

3. For the next 3 weeks, complete general monitoring once per week to restock HFD, weigh animals, and assess non-fasting blood glucose levels. Assessment of non-fasting blood glucose should be done at approximately the same time of day throughout the protocol to avoid the impact of the circadian cycle.

### 3.2 Osmotic Pump Preparation

Osmotic pumps should be prepared on the day of the implant. Pre-weigh animals to determine drug calculations.

1. Lay out sterile sheets and all required equipment (*see* Subheading 2.2, **items 2–5**). Use this as a sterile field to prepare STZ and mini pumps.
2. Prepare 150 μL of STZ in citrate-buffered saline. Vortex until fully dissolved (*see* **Note** 7).
3. Connect the blunt, 27G needle (provided in the osmotic mini pump pack) to a 1 mL syringe, and draw up STZ solution ensuring there are no air bubbles.
4. Insert the blunt needle all the way into the pump reservoir allowing the needle to rest at the base of the reservoir. Slowly fill the reservoir from the bottom to the top while holding the pump in a vertical orientation (*see* **Note 8**). When the pump is completely full, a small liquid bubble will overflow from the reservoir where the needle is inserted.
5. Remove the needle and insert the flow regulator into the reservoir (*see* **Note 9**).
6. Store the mini pump in sterile saline at 37 °C until required (*see* **Note 2**).

### 3.3 Osmotic Mini Pump Implant and Removal

Osmotic mini pumps containing STZ should be implanted in week 3.

1. Lay out sterile sheets and all required sterile equipment (*see* Subheading 2.2, **items 9–11**). Prepare a syringe containing meloxicam.
2. Anesthetize in the induction box using 5% isoflurane carried in oxygen.
3. Confirm appropriate anesthesia by monitoring movement and respiratory rate (the mouse should have no conscious movement and have a respiratory rate of approximately 60 breaths/min), relocate the mouse to the nose cone, and maintain anesthesia

using 2–3% isoflurane in oxygen as required. Before proceeding, assess the level of anesthesia using a strong tail or foot pinch—if a reflex is observed, adjust the concentration of isoflurane to ensure the mouse is maintained under a deep level of anesthesia.

4. Inject subcutaneously with meloxicam below the lumbar region. Use forceps to lift the skin and insert the needle into the subcutaneous space and inject.
5. Shave an area (~2 cm × 2 cm) on the dorsum at the cervical/thoracic level, and sterilize the skin using Betadine solution and a cotton tip applicator.
6. Using surgical scissors, make a 1 cm incision (perpendicular to the spine and above the scapulae) through the skin.
7. Insert the hemostat under the skin toward the tail. By opening and closing the hemostat, create a 3–4-cm-long pocket between the skin and muscle tissue to fit the osmotic mini pump in the subcutaneous space.
8. Insert the osmotic mini pump into the subcutaneous pocket, regulator end first. It is important to ensure delivery of the solution is at the opposite end to the incision site (*see* Fig. 1).
9. Pucker the skin on either side of the incision using forceps, and seal using two wound clips (*see* **Notes 10** and **11**).
10. House mice individually for 24 h to recover, and then return to original housing arrangements (*see* **Note 12**).
11. It is recommended that non-fasting blood glucose and body weight are assessed daily following osmotic mini pump implantation (*see* **Note 13**).
12. This model of osmotic mini pump delivers the desired solution across 14 days. Immediately after the 5-week point, osmotic mini pumps should be removed. For this, repeat **steps 1–5**, and then make a 1 cm incision just below the insertion scar. Using surgical scissors, dissect any scar tissue formed, and extract the osmotic mini pump. Complete **steps 9** and **10**.

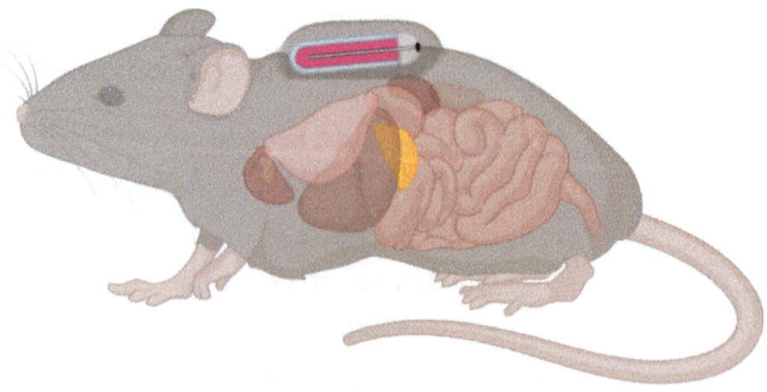

**Fig. 1** Osmotic mini pump is placed subcutaneously on the animal's back. The pump should be positioned below the scapulae with the regulator directed toward the tail. (Figure made using Biorender.com)

### 3.4 Glucose Tolerance Testing

Glucose tolerance tests (GTT) should be performed at the same time of day across this protocol (e.g., 1 pm) to reduce the impact of the circadian cycle.

1. All mice will need to be fasted for 6 h by placing animals into new cages without food at 7 am (*see* **Note 14**).
2. At approximately 12:30 pm, weigh mice, and calculate the required volume of 30% glucose required to achieve a 2.0 g/kg dose. For example, if a mouse is 30 g, then 0.03 kg × 2.0 g/kg = 0.06 g glucose. A 30% glucose solution contains 0.3 g glucose/mL. Therefore, 0.06 g ÷ 0.3 g/mL = 0.2 mL. A 30 g mouse needs 0.2 mL of 30% glucose solution for a dose of 2.0 g/kg.
3. Load glucose solution into a 0.5 mL syringe and set aside.
4. Cover the mouse tail tip in topical anesthetic cream and leave for 5 min.
5. Insert glucose test strip into glucose monitor. Leaving the mouse in the home cage, make a small cut (≤1.0 mm) to remove the tip of the tail. Collect the drop of blood on the test strip, and record blood glucose concentration. This is the baseline glucose recording (0 min) and represents the animal's fasting blood glucose concentration.
6. At 1 pm, restrain the mouse and inject intraperitoneally with a prepared glucose solution.
7. Start timer and repeat glucose sampling at 10, 20, 30, 45, 60, 75, 90, and 120 min post-glucose injection. The initial tail cut should be sufficient for regular blood sampling by running your fingers along the length of the tail (*see* **Note 15**).
8. At the end of the sampling period, return animals to their original cages continuing HFD.
9. The GTT can be repeated multiple times across the course of the experiment to track changes in glucose tolerance over time following treatments (*see* Fig. 2) [5].

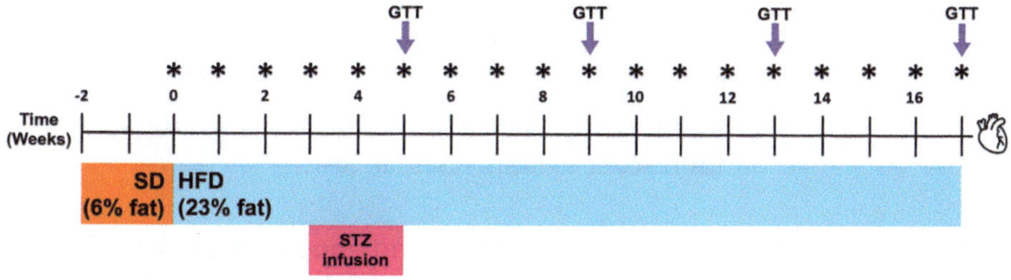


**Fig. 2** A 17-week timeline for inducing type 2 diabetes using high-fat diet (HFD) and streptozotocin (STZ). Animals were maintained on a standard laboratory diet (SD) prior to day 0. Asterisk indicates weekly checking of body weight and blood glucose. Arrows indicate when glucose tolerance tests (GTT) were performed. After week 17, animals were transcardially perfused to generate a fluorescent vascular cast

### 3.5 Cardiac Perfusion with Fluorescent Gelatin Vascular Casting

1. Prepare the peristaltic pump line by attaching a 23G needle to the end and ensuring a flow rate of 7.5–8.0 mL/min. Prefill line with 1x heparinized PBS (*see* **Note 16**).
2. Restrain the mouse and inject intraperitoneally with 180 mg/kg pentobarbitone for deep, terminal anesthesia.
3. Check the level of anesthesia by pinching the toe or tail with forceps. When the pain reflex is gone, use surgical scissors to make a vertical 3 cm incision through the skin of the abdomen just below the rib cage to expose the abdominal cavity, sternum, and diaphragm.
4. Pierce the diaphragm and cut through the ribs on either side of the chest cavity ensuring not to damage the lungs or heart.
5. Holding the sternum with forceps, make a lateral cut disconnecting the diaphragm from the rib cage (*see* **Note 17**).
6. Clamp the sternum with the hemostat, and fold upward over the head exposing the full chest cavity.
7. Gently remove connective tissue surrounding the heart. Gently hold the apex of the heart with forceps, and insert the blunt 26G needle into the left ventricle. It is critical to ensure the needle is in the left ventricle so that fluid moves properly through the systemic circulation. Turn the pump on for 1 s, and stop to ensure the needle is correctly positioned in the left ventricle and there are no leaks (*see* **Note 18**).
8. Keeping the needle in position, make a small cut in the right atrium allowing blood to leak from the right side of the heart as the blood is flushed from the vascular system.
9. Start the pump and perfuse with approximately 10 mL of heparinized PBS.
10. Stop the pump and change the reservoir to 4% PFA (*see* **Note 19**). Restart the pump and perfuse with approximately 10 mL PFA.
11. Stop the pump, move the infusion line to the FITC-dextran/gelatin solution, and perfuse with approximately 10 mL (*see* **Note 20**).
12. Remove the needle from the ventricle, and immediately flush the perfusion line with warm water to prevent gelatin setting and clogging the infusion line for the next mouse.
13. Submerge the body of the mouse in ice for 20 min to set the fluorescent gelatin vascular cast.

### 3.6 Retinal Dissection

The retinas are dissected and mounted as described previously [10]. Briefly, remove the top of the skull and brain to access and remove the eyes. Dissect any accessory muscles and connective tissue still attached. Remove the cornea and lens leaving behind

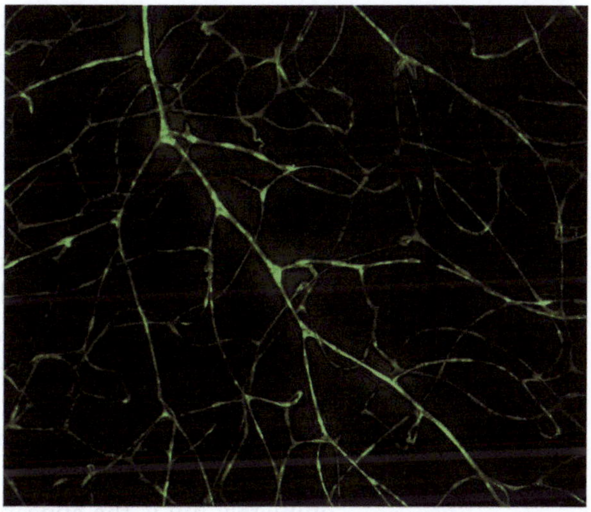

**Fig. 3** Example image of a fluorescent gelatin cast of the retinal vasculature. The image shows a small section of the retina where blood vessels have been filled with gelatin containing 0.5% FITC-dextran. Scale bar 50 μm

the eye cup and retina. Remove the sclera leaving behind the cup-shaped retina. Partially cut the retina into quarters, leaving the quarters attached in the center. Mount on a microscope slide and image using a fluorescence microscope (*see* Fig. 3).

## 4 Notes

1. A dose of 200 mg/kg STZ was the dose that generated moderate hyperglycemia in this strain of mouse. The exact dose required to achieve a similar effect in other strains of mice will likely vary slightly and need to be determined by individual groups/researchers.
2. We obtained the most consistent results when osmotic mini pumps were loaded with STZ immediately before implant.
3. We recommend making enough 30% glucose solution to perform all GTTs across the duration of the procedure. This will help to minimize variation in GTT as a result of variable glucose doses. 30% glucose solution can be stored in the freezer for >12 months.
4. We have found that the process of dissolving 3 g D-glucose in 7 mL saline can be time-consuming. Heating the saline to boiling point (e.g., using a microwave) speeds up this process.
5. Surgical tools that are used to perform cardiac perfusions should not be used for in vivo experiments due to contamination with PFA. We suggest having a set of surgical tools used exclusively for cardiac perfusions and dissections.

6. As FITC-dextran is a carbohydrate conjugated to a 488 nm fluorophore, it is important to keep the solution in the dark to prevent bleaching. We recommend wrapping the tube containing the solution in foil and limiting light exposure as much as possible.
7. If the STZ fails to readily dissolve, we recommend checking that the citrate-buffered saline has a pH 4.4. It is important not to load mini pumps where particulates remain in the solution as this may block the flow regulator preventing STZ infusion.
8. It is important that the osmotic mini pump reservoir is filled relatively slowly to avoid the formation of air bubbles. Small air bubbles will impact the volume of solution that the reservoir can hold and can impair mini pump performance.
9. A good indicator that the osmotic mini pump reservoir is completely full is the appearance of a small liquid bubble out of the delivery portal as the regulator is being inserted.
10. The subcutaneous pocket where the osmotic mini pump is implanted needs to be large enough that the mini pump sits approximately 1 cm below the incision site. This reduces pressure on the wound and favors better wound healing.
11. To ensure the wound clips are securely in place and reduce the likelihood of them falling out prematurely, clamp them tightly with a hemostat. Additionally, we recommend using in-cage housing with smooth edges as we found animals would get wound clips caught on rougher surfaces.
12. We suggest that animals be housed individually for no longer than 24 h. Housing animals individually for extended lengths of time causes distress and impacts glucose homeostasis [11].
13. As a guide, if non-fasting blood glucose levels exceed 25 mmol/L during the STZ infusion period, this indicates total beta-cell loss and transition to type 1 diabetes. From our experience, this occurs in $\leq$5% of mice and is likely caused by different sensitivity to STZ in some mice. In these instances, mini pumps can be removed on the day the blood glucose levels reach this point, and each individual animal's non-fasting glucose should be monitored daily. Animals that continue to have extremely high non-fasting blood glucose ($\geq$25 mmol/L) for more than 5 days after mini pump removal should be removed from the experiment as they do not represent the type 2 diabetic model accurately.
14. We have observed mice rapidly shredding the HFD food and piling it in the cage. To ensure accurate fasting prior to GTTs, place animals into new cages rather than just removing the food hopper.

15. If the tail stops bleeding across the 120 min GTT, use a damp Kimwipe to remove the clotted blood. If this is insufficient, a second cut can be made. If bleeding persists beyond the 120 min GTT, we recommend using a tissue adhesive such as Vetbond to seal the wound.
16. Keeping all solutions in 50 mL falcon tubes means they can be arranged in the correct order (PBS, PFA, and then gel) in a tube rack, with lids removed to ensure a quick transition between each solution.
17. We recommend slowly cutting through the diaphragm as often the heart is positioned within connective tissue close to the diaphragm. Any damage to the heart can impact the quality of the gelatin vascular cast.
18. Aim to commence cardiac perfusion while the heart is still beating. If the heart stops before perfusion begins, hemostasis may occur, and blood will begin to coagulate and may impact upon PFA and gelatin perfusion.
19. Stopping the pump when changing solutions helps to prevent bubbles appearing in the line which would impair fluid flow in the vasculature and gelatin distribution.
20. When completing multiple perfusions in succession, we recommend keeping the gelatin solution in a 37 °C water bath between procedures to prevent the gelatin setting prematurely.

## Acknowledgments

This work was supported by the University of Tasmania Research Enhancement Program grant.

### References

1. Roglic G (2016) WHO global report on diabetes: a summary. Int J Noncommun Dis 1(1):3
2. Leasher JL, Bourne RR, Flaxman SR, Jonas JB, Keeffe J, Naidoo K, Pesudovs K, Price H, White RA, Wong TY (2016) Global estimates on the number of people blind or visually impaired by diabetic retinopathy: a meta-analysis from 1990 to 2010. Diabetes Care 39(9):1643–1649
3. Daniel A, Premilovac D, Foa L, Feng Z, Shah K, Zhang Q, Woolley KL, Bye N, Smith JA, Gueven N (2021) Novel short-chain quinones to treat vision loss in a rat model of diabetic retinopathy. Int J Mol Sci 22(3):1016
4. Premilovac D, Gasperini RJ, Sawyer S, West A, Keske MA, Taylor BV, Foa L (2017) A new method for targeted and sustained induction of type 2 diabetes in rodents. Sci Rep 7(1):14158
5. Southam K, Sousa C, Daniel A, Taylor BV, Foa L, Premilovac D (2022) Development and characterisation of a rat model that exhibits both metabolic dysfunction and neurodegeneration seen in type 2 diabetes. J Physiol 600(7):1611–1630. https://doi.org/10.1113/jp282454
6. Schnedl WJ, Ferber S, Johnson JH, Newgard CB (1994) STZ transport and cytotoxicity: specific enhancement in GLUT2-expressing cells. Diabetes 43(11):1326–1333. https://doi.org/10.2337/diab.43.11.1326
7. Yau JWY, Rogers SL, Kawasaki R, Lamoureux EL, Kowalski JW, Bek T, Chen S-J, Dekker JM, Fletcher A, Grauslund J, Haffner S, Hamman RF, Ikram MK, Kayama T, Klein BEK, Klein R, Krishnaiah S, Mayurasakorn K, O'Hare JP, Orchard TJ, Porta M, Rema M, Roy MS, Sharma T, Shaw J, Taylor H, Tielsch JM,

Varma R, Wang JJ, Wang N, West S, Xu L, Yasuda M, Zhang X, Mitchell P, Wong TY (2012) Global prevalence and major risk factors of diabetic retinopathy. Diabetes Care 35(3):556–564. https://doi.org/10.2337/dc11-1909

8. Papanas N, Ziegler D (2015) Risk factors and comorbidities in diabetic neuropathy: an update 2015. Rev Diabet Stud 12(1–2):48

9. Dronavalli S, Duka I, Bakris GL (2008) The pathogenesis of diabetic nephropathy. Nat Clin Pract Endocrinol Metab 4(8):444–452

10. Claybon A, Bishop AJR (2011) Dissection of a mouse eye for a whole mount of the retinal pigment epithelium. J Vis Exp (48). https://doi.org/10.3791/2563

11. Arndt SS, Laarakker MC, van Lith HA, van der Staay FJ, Gieling E, Salomons AR, van't Klooster J, Ohl F (2009) Individual housing of mice—impact on behaviour and stress responses. Physiol Behav 97(3–4):385–393

# Chapter 2

# Akimba Proliferative Diabetic Retinopathy Model: Understanding Molecular Mechanism and Drug Screening for the Progression of Diabetic Retinopathy

**Sai Bo Bo Tun and Veluchamy Amutha Barathi**


## Abstract

As the prevalence of diabetes has reached epidemic proportions worldwide, diabetic retinopathy incidence is increasing rapidly. An advanced diabetic retinopathy (DR) stage can lead to a sight-threatening form. There is growing evidence showing diabetes causes a range of metabolic changes that subsequently lead to pathological modifications in the retina and retinal blood vessels. To understand the complex mechanism of the pathophysiology of DR, a precise model is not readily available. By crossbreeding the Akita and Kimba strains, a suitable proliferative DR model was acquired. This new Akimba strain manifests marked hyperglycemia and vascular changes, which resemble the early and advanced stage of DR.

Here, we describe the breeding method, colony screening for experiments, and imaging techniques widely used to investigate the DR progression in this model. We elaborate step-by-step protocols to set up and perform fundus, fluorescein angiography, optical coherence tomography, and optical coherence tomography-angiogram to study retinal structural changes and vascular abnormalities. In addition, we show a method to label the leukocytes with fluorescence and laser speckle flowgraphy to examine the inflammation in the retina and retinal vessel blood flow speed, respectively. Lastly, we describe electroretinogram to evaluate the functional aspect of the DR transformations.


**Key words** Akimba, Diabetic retinopathy, Proliferative diabetic retinopathy, Murine diabetic model, Ocular imaging

## 1 Introduction

Diabetic retinopathy (DR) is one of the most common eye diseases and a leading cause of blindness as the incidence of diabetes is increasing dramatically. Currently, 537 million adults have diabetes, and one-third of these patients are suffering from diabetic retinopathy [1]. The pathophysiology of DR is mainly due to poor glycemic control leading to the production of advanced glycation end products, which eventually induce pathological changes in the retina. DR is a chronic progressive disease with vascular changes and capillary loss at an early stage, followed by alterations in the

Guei-Sheung Liu and Jiang-Hui Wang (eds.), *Diabetic Retinopathy: Methods and Protocols*, Methods in Molecular Biology, vol. 2678, https://doi.org/10.1007/978-1-0716-3255-0_2,

retinal blood vessel integrity and neuronal layers. Therefore, it is crucial for an early diagnosis of diabetes and DR to prevent vision loss. However, the complex molecular mechanism of progression of DR has not been fully described yet despite several years of effort. Hence, an animal model that can mimic the pathophysiology of DR is an asset for the evaluation of underlying mechanisms of DR and the development of novel therapies.

There is no precise animal model for DR, especially in rodents. Most models of diabetes-induced eye disease, such as streptozotocin-induced rats and mice, *Ins2$^{Akita}$* mice, and db/db (*Lepr$^{db}$*) mice, show early retinopathy changes, while progressive changes need to be induced with "on top" alternative approaches such as laser photocoagulation [2, 3]. An accurate model to study different stages of DR is not readily available. The Akimba mouse is the first animal model presenting hyperglycemia and advanced diabetes-related complications in both the eye and kidney from young. Akimba (*Ins2$^{Akita}$VEGF$^{+/-}$*) is a novel strain generated by crossing the heterozygous Akita (*Ins2$^{Akita}$*) line and homozygous Kimba (*VEGF$^{+/+}$*) line [4]. The Akimba line has been used for several experimental approaches to study DR pathophysiological changes [5–7] and a therapeutic strategy for DR [8–11].

The Akita (*Ins2$^{Akita}$*) mouse is a non-obese mutant mouse with spontaneous early age-onset diabetes, marked hyperglycemia throughout their life, elevated blood pressure, polydipsia, and polyuria. The point mutation in *Ins2$^{Akita}$* mice leads to conformational changes in insulin protein, rendering spontaneous diabetes with a significant neuronal loss at the inner nuclear layer (INL) and inner plexiform layer (IPL) of the retina [12]. It is regarded as an excellent model for studying the mechanism of early DR changes as it is associated with acellular capillaries, pericyte and endothelial cell loss, leukostasis, and increased microglial activity. After 22 weeks of hyperglycemia, there was a thinning of inner retinal layers suggesting chronic degeneration [12]. In addition, there was a reduction in the number of cholinergic and dopaminergic amacrine cells in this strain of mouse [13]. However, there was no vascular leakage or prominent neovascularization, signs of proliferative changes of advanced DR.

Researchers have been using hVEGF overexpressed mouse strain to study proliferative changes in the retina. The Kimba (*VEGF$^{+/+}$*) mice do not possess hyperglycemia; nevertheless, age-dependent retinal vascular changes are observed. The photoreceptors of Kimba mice overexpress hVEGF transiently during early age, which leads to vascular pathologies such as increased vascular permeability, microaneurysms, capillary dropout, vessel tortuosity, leukostasis, and prominent retinal neovascularization [4, 14]. These vascular changes directly resemble the phenotypes of advanced DR and showed progression with age [15]. The neural retinal thickness of the Kimba mice was also significantly lower than the C57BL/6 mice [4]. The Kimba mouse model is suitable for testing anti-angiogenic molecules designed to target hVEGF.

The Akimba ($Ins2^{Akita}VEGF^{+/-}$) mice demonstrate features exhibited by both Kimba and Akita mice. These include (i) onset of hyperglycemia from 4 weeks post-natal with consistent high plasma glucose levels throughout their life, (ii) retinopathic changes exhibited by the Kimba mice, (iii) nephropathic changes similar to those present in the Akita mice, (iv) normal weight gain through 18 weeks of age followed by no further weight gain and subsequent weight loss at 30 weeks old, and (v) loss of adipose tissues with age [4–6]. The Akimba mice are suitable to study diabetes-related retinal complications as this line expresses both metabolic changes and vascular complications. It includes vessel constriction, capillary leakage, capillary non-perfusion, beading, and obvious retinal neovascularization accompanied by hyperglycemia. In addition, there was an age-related ganglion cell loss in the Akimba mice. However, the vascular changes in the Akimba mice are not due to prolonged hyperglycemic attack but rather because of hVEGF overexpression. This progressive Akimba model would enable us to characterize the structural and functional changes of DR. This model is also useful for investigating novel pharmacological interventions to prevent microvascular changes in diabetes-induced retinal vascular inflammation, edema, and angiogenesis.

Here, we described several methods used in this model to evaluate the genotyping, phenotype characterization, pathophysiology of disease progression, and testing of novel pharmacological compounds in treating or preventing DR. Retinal fundus imaging reveals phenotypical changes in the retina and vessels, such as neovascularization, hemorrhage, and morphological changes of the retina [4, 16]. The fundus fluorescein angiography (FFA) shows discernible retinal vasculature, capillaries, quantifiable vascular leakage from the unhealthy blood vessels, and retinal/choroidal neovascularization [4, 5, 17]. Optical coherence tomography (OCT) and optical coherence tomography-angiogram (OCT-A) provide various information such as cross-sectional morphological images of retinal layers and choroid, loss of retinal layers, retinal edema, and vascular changes in both the retina and choroid [5, 18–21]. Laser speckle flowgraphy (LSFG) can be used to measure the vessel diameter and relative blood flow velocity in the retinal blood vessels [5, 22, 23]. Leukostasis can be detected by using the fluorescein-labelled leukocyte technique, revealing the vascular inflammation and infiltrating leukocytes in the retina [24, 25]. Electroretinogram (full-field ERG) is an essential technique to evaluate the retinal function. The physiological and pathological changes of the retinal cell functions can be monitored overtime on the same animal with or without treatment. The a-wave reflects the general physiological health of the photoreceptors (rods and cones) in the outer retina. In contrast, the b-wave reflects the health of the inner retinal layers, including the ON bipolar cells and the Muller cells. In addition, specific rod and cone functions, retinal neuron function, and ganglion cell function loss can be measured by performing in the dark- and light-adapted conditions [5, 26].

## 2 Materials

### 2.1 Breeding and Screening of Akimba Mice

1. C57BL/6 J heterozygote Akita (*Ins2*$^{Akita}$) mice (Jackson Lab, USA).
2. Homozygote Kimba (*VEGF*$^{+/+}$) (Lions Eye Institute, Perth).
3. Disposable 27G needle.
4. Alcohol swab.
5. Handheld glucometer.

### 2.2 Genotyping

1. TaqMan® Sample-to-SNP™ kit.
2. TaqMan® GTXpress™ Master Mix.
3. TaqMan® genotyping assay mix: Probe sequence (Akita-Fw, CAG ACC TTG GCA CTG GAG GT; Akita-Rev, CAG AGG GGT AGG CTG GGT AGT; Akita-p1-VIC, VIC-GAT CAG TGC TGC ACC; Akita-p2-FAM, 6FAM-GTA GAT CAG TGC TAC ACC).
4. 1× TE buffer: 10 mM Tris-HCl, 1 mM EDTA, pH 8.0.
5. 1.5 mL microcentrifuge tubes.
6. Sterile scissor.
7. Incubator with shaking function.
8. Real-time PCR machine.

### 2.3 Preparation of the Animals for In Vivo Imaging

1. Ketamine.
2. Xylazine.
3. 1% tropicamide.
4. 1% phenylephrine.
5. Vidisic® ophthalmic gel.
6. Warm light.

### 2.4 Fundus and Fundus Fluorescein Angiography (FFA)

1. Vidisic® ophthalmic gel.
2. 10% sodium fluorescein.
3. 1 mL tuberculin syringe with 27G needle.
4. MICRON® IV rodent eye system (Phoenix-Micron, Inc., Bend, Oregon, USA).
5. Warm light.

### 2.5 Optical Coherence Tomography (OCT)

1. Vidisic® ophthalmic gel.
2. Image-guided OCT on a MICRON® IV rodent eye system (Phoenix-Micron, Inc., Bend, Oregon, USA).
3. Warm light.

### 2.6 Optical Coherence Tomography-Angiogram (OCT-A)

1. XYZ translational head holder.
2. Vidisic® ophthalmic gel.
3. Contact lens designed for mouse (refractive index, 1.51; radius of curvature, 1.70 mm; diameter, 3.2 mm).
4. ZEISS Swept-Source-OCTA Plex® Elite 9000 system (Zeiss, Carl Zeiss Meditec, Dublin, California, USA).
5. Warm light.

### 2.7 Laser Speckle Flowgraphy (LSFG)

1. Vidisic® ophthalmic gel.
2. Round coverslip (diameter: 12 mm).
3. LSFG®-Micro machine (Softcare Co., Ltd., Fukuoka, Japan).
4. Warm light.

### 2.8 Fluorescein-Labelled Leukocytes to Detect Leukostasis

1. 1 mL tuberculin syringe with 27G needle.
2. EDTA tubes (EDTA K3 blood collection tubes).
3. Histopaque®-1077 gradient solution: 1.077 g/mL polysucrose.
4. 1× PBS.
5. 10% sodium fluorescein.
6. Vidisic® ophthalmic gel.
7. Contact lens designed for mouse (refractive index, 1.51; radius of curvature; 1.70 mm; diameter, 3.2 mm).
8. Insulin syringe attached to a 31G needle.
9. 70% isopropyl alcohol swab.
10. Temperature-controlled centrifuge.
11. Spectralis® HRA+OCT machine (Heidelberg Engineering, Heidelberg, Germany).
12. Warm light.

### 2.9 Electroretinogram (ERG)

1. Gold wire cornea electrode (2 mm for baby mouse, 3 mm for adult mouse).
2. Gold cup electrode.
3. Platinum hypodermic needle electrode.
4. Red light head lamp.
5. Espion® Diagnosys system with a ColorDome™ LabCradle™ (Diagnosys LLC, Redwood City, CA, USA).

## 3 Methods

### 3.1 Breeding and Screening of Akimba Mice

1. Crossbreed the two strains, i.e., Akita (*Ins2$^{Akita}$*) and Kimba (*VEGF$^{+/+}$*), to get Akimba (*Ins2$^{Akita}$ VEGF$^{+/-}$*) double strain (*see* **Note 1**).
2. Screen the colony by measuring the fasting plasma glucose level of the mice using a handheld glucometer. Remove the feed from the cage 6 h before the plasma glucose measurement. Water is allowed at ad libitum. Restrain the mouse in a restrainer with the tail exposed. Clean the tip of the tail with a 70% isopropyl alcohol swab, and air-dry the surface (*see* **Note 2**). Prepare a glucometer and a test strip to receive the blood sample. Use a disposable 27G needle to prick the tip of the tail, and squeeze out a drop of blood. Apply the blood to the collection tip of the glucometer strip. Choose the animals with a glucose level of more than 20 mmol/L.
3. Monitor plasma glucose level of the mice using a handheld glucometer every 3 weeks starting from 2 weeks old of age and fluorescein angiography weekly from 4 weeks of age to select Akimba mice for experiments. Akimba mice are hyperglycemic together with retinal vessel leakage from 4 weeks of age, and the disease progresses extensively after 6–8 weeks of age.
4. Use confirmed Akimba (*Ins2$^{Akita}$ VEGF$^{+/-}$*) double strain for further experiments, and maintain Akita (*Ins2$^{Akita}$*) x Kimba (*VEGF$^{+/+}$*) to further expand the colony to get Akimba (*Ins2$^{Akita}$ VEGF$^{+/-}$*) double strain.

### 3.2 Genotyping

1. At 1–2 weeks old, collect 2–3 mm of tail biopsy using a sterile scissor to extract DNA for genotyping. Restrain the mouse in a restrainer, and sterilize the tip of the tail by rubbing it with an alcohol swab. Air-dry the alcohol and cut the tail 2–3 mm from the end (*see* **Note 3**). Transfer the tip clip into a new sterile 1.5 mL microcentrifuge tube. Store the samples in a −80 °C freezer until ready for digestion.
2. Prepare the TaqMan® Sample-to-SNP™ kit according to the manufacturer's instructions. Bring out the tail samples from the freezer and keep them on ice.
3. Thoroughly mix the lysis solution and add 50 μL of the solution into the sample containing tubes. Mix the lysis solution and the tail samples well by pipetting up and down. Cap the tubes and run a quick centrifuge.
4. Incubate the tail samples at 95 °C for 3 min. Cool down the tubes at room temperature for 30 s.

5. Thoroughly mix the DNA stabilizing solution and add 50 μL of the solution into the sample containing tubes. Mix the solution and the samples well by pipetting up and down. Cap the tubes and run a quick centrifuge. Store the crude lysates at −4 °C for a short-term (up to 1 week) or − 20 °C for long-term storage.
6. Mix the TaqMan® GTXpress™ Master Mix thoroughly without creating bubbles. Thaw on ice, vortex, and centrifuge the TaqMan® assay reagents and samples. For each sample, prepare the PCR reaction mix by combining 5 μL of TaqMan® GTXpress Master Mix (2×) and 0.5 μL of TaqMan® genotyping assay mix (20×) and 2.5 μL of DNase-free water (*see* **Note 4**).
7. Cap and vortex the tubes followed by a brief spinning down of the content. Pipette 8 μL of PCR reaction mixes into each well of a reaction plate (*see* **Note 5**) and sealed with a MicroAmp™ adhesive film. Centrifuge the plate briefly to eliminate the air bubbles.
8. Add 2 μL of sample lysate and control and cover the plate with MicroAmp™ adhesive film. Centrifuge the plate briefly to eliminate the air bubbles.
9. Set up the PCR according to the following run conditions.

| Stage | Step | Temperature | Time |
|---|---|---|---|
| Holding | DNA polymerase activation | 95 °C | 20 s |
| Cycling (40 cycles) | Denature<br>Anneal/extend | 95 °C<br>60 °C | 3 s<br>20 s |

10. Load the reaction plate into the thermal cycler, and start the run selecting a run speed as standard and a reaction volume as 10 μL. Read the result on a real-time PCR instrument according to the manufacturer's instructions.

### 3.3 Preparation of the Animals for In Vivo Imaging

1. Anesthetize the mouse with a combination of ketamine at 20 mg/kg body weight and xylazine at 2 mg/kg body weight.
2. After administering anesthesia, administer a drop each of 1% tropicamide and 2.5% phenylephrine to dilate the pupil (*see* **Note 6**).
3. Apply a small amount of Vidisic® ophthalmic gel to the cornea, and wait for 5 min for full pupil dilation (*see* **Note 7**).

4. Place the mouse in a warm empty cage until it is fully anesthetized. Test the level of anesthesia by pinching the paw pad with the thumb and index finger. When there is no response to pinching, proceed to respective imaging.

### 3.4 Fundus and Fundus Fluorescein Angiography (FFA)

1. Acquire fundus photography and fundus fluorescein angiography using a Phoenix MICRON® IV rodent system.
2. Turn on the MICRON® machine according to the manufacturer's instructions, and prepare the mouse as described in Subheading 3.3.
3. Place the mouse on the stage in front of the machine with the eye looking straight toward the objective. Apply a small amount of Vidisic® ophthalmic gel to the eye. Turn on the light of the MICRON® machine, and slowly move the objective closer to the eye until the objective touches the gel. Make sure the pupil is aligned at the center of the view, and slowly advance the objective. Stop advancing the objective when the whole fundus picture of the eye is in the frame (*see* **Note 8**).
4. Adjust the brightness of the light, and focus to acquire sharp and well-exposed fundus images of the mouse. Acquire the images by pressing the foot switch attached to the machine.
5. Prepare 10% sodium fluorescein in a 1 mL tuberculin syringe. Inject the fluorescein intraperitoneally at 0.01 mL/6 g body weight. Align the mouse again as described in **step 3** and turn on the blue filter. Adjust the exposure time and focus to get sharp images. Acquire the angiograms at immediately, 3, 7, and 10 min after the injection.
6. Place the mouse under a warm light to recover (*see* **Note 9**).

### 3.5 Optical Coherence Tomography (OCT)

1. Acquire the OCT images of the retina using an image-guided OCT on a Phoenix MICRON® IV rodent system (*see* **Note 10**).
2. Turn on the MICRON® OCT system as per the manufacturer's instructions, and prepare the mouse as described in Subheading 3.3.
3. Position the eye of the mouse to get a clear fundus image as described in Subheading 3.4, **step 3**. Move the scanning line indicator on the fundus to the area of interest. Adjust the reference arm of the OCT to get a full cross-sectional OCT image of the retina. Take an average of a stack of 20 images (*see* **Note 11**). Save the acquired OCT images.
4. Place the mouse under a warm light to recover.

### 3.6 Optical Coherence Tomography-Angiogram (OCT-A)

1. Acquire OCT-A images using a ZEISS Swept-Source-OCTA Plex® Elite 9000 system.
2. Turn on the Plex® Elite as per the manufacturer's instructions, and prepare the mouse as described in Subheading 3.3.
3. Place the mouse on the stage with an XYZ translational head holder in front of the Plex® machine (*see* **Note 12**). Apply a small amount of Vidisic® ophthalmic gel and a contact lens designed for the mouse on the cornea. Position the mouse eye looking straight toward the objective, and slowly move the stage nearer to the machine. Adjust the focus of the machine to get a clear fundus image, and angle the retina appropriately (*see* **Note 13**).
4. Use a configuration of $12 \times 12$ mm$^2$ scanning protocol (*see* **Note 14**). Start the scanning and save the OCTA images.
5. Place the mouse under a warm light to recover.

### 3.7 Laser Speckle Flowgraphy (LSFG)

1. Obtain the blood flow velocity measurement by using a LSFG®-Micro machine.
2. Turn on the LSFG®-Micro system according to the manufacturer's instructions, and prepare the mouse as described in Subheading 3.3.
3. Put the mouse on a pillow on the imaging stage with the eye looking straight upward under the scanning objective. Apply a copious amount of Vidisic® ophthalmic gel to cover the whole eye, and place a round coverslip on top of the gel. This will allow you to visualize the mouse retina (*see* **Note 15**). Adjust the power of the laser in the system to get an optimal exposure.
4. By moving the position of the microscope objective, position the optic head at the center of the eye. Fix the position of the microscope, and adjust the focus by moving the Z position of the microscope to get a sharp vessel outline.
5. Acquire the LSFG images and save the images.
6. Place the mouse under a warm light to recover.

### 3.8 Fluorescein-Labelled Leukocytes to Detect Leukostasis

1. Anesthetize the mouse with a combination of ketamine at 20 mg/kg body weight and xylazine at 2 mg/kg body weight. Test the level of anesthesia by the toe-pinching method described in Subheading 3.3.
2. Place the mouse in a supine position, and identify the inferior end of the sternum. Insert a 27G needle attached to a 1 mL tuberculin syringe pointing directly toward the heart, and advance slowly. Withdraw the plunger slowly while advancing the needle. Once the blood is seen flowing into the syringe,

stop the advancement. Gently withdraw the plunger until 1 mL of blood is collected. Withdraw the needle and the syringe from the mouse.

3. Remove the needle from the syringe, and transfer the whole blood into an ethylenediaminetetraacetic acid (EDTA) tube (*see* **Note 16**). Gently roll the tube to mix the blood and the EDTA in the tube. Keep the exsanguinated blood on ice until processing. Euthanize the mouse with a pentobarbital overdose.
4. Prepare a 2 mL microcentrifuge tube with mL of 1.077 g/mL Histopaque® gradient solution polysucrose solution. By using a 1 mL micropipette, gently layer the collected whole blood on top of the gradient solution, and centrifuge at 450×*g* at 4 °C for 30 min (*see* **Note 17**). Discard the supernatant (plasma) using a pipette, and transfer the buffy coat (leukocytes) into a new tube.
5. Wash the leukocytes with 10 mL of 1× PBS, and centrifuge at 45×*g* at 4 °C for 10 min.
6. Resuspend the cell pellet in 900 μL of 1× PBS. Add 100 μL of 10% sodium fluorescein into the tube, and mix with the cells thoroughly. Incubate at room temperature for 2 min in the dark.
7. Wash the labelled leukocytes three times in 10 mL of 1× PBS, and centrifuge at 450×*g* at 4 °C for 10 min.
8. Resuspend the labelled cells in 100 μL of 1× PBS. Wrap the tube containing the labelled cells in an aluminum foil to prevent photobleaching (*see* **Note 18**).
9. In vivo imaging was performed by using a Spectralis® HRA +OCT machine (Heidelberg, Germany) with a 25 diopter lens fitted on a 30-angle lens.
10. Turn on the Spectralis® machine according to the manufacturer's instructions. Prepare the mouse as described in Subheading 3.3. Keep the mouse in a warmed cage after the anesthesia until ready for a tail vein injection (*see* **Note 19**).
11. Apply a small amount of Vidisic® ophthalmic gel to the cornea, and put a contact lens designed for mice on the cornea. Place the mouse in front of the objective, and start a high-speed imaging mode on the Spectralis®.
12. Withdraw the labelled cells into an insulin syringe attached to a 31G needle. Sterilize the mouse tail using a 70% isopropyl alcohol swab, and inject the labelled cells via a tail vein injection.
13. Using infrared (IR) imaging, position the optic head at the center of the eye by moving the Spectralis® imaging module.

Turn on the fluorescence angiography mode with an excitation light provided by a 488 nm argon laser. The moving fluorescence-labelled leukocytes can be seen in the retinal vessels of the mouse.

14. Take 30 s videos immediately, 5 min, and 15 min after labelled cell injection.
15. Place the mouse under a warm light to recover.

### 3.9 Full-field Electroretinogram (ffERG)

1. Acquire ffERG using an Espion® Diagnosys system with a ColorDome™ LabCradle™ setup. Acquire all ERG records at 2 kHz with high- and low-pass filters set to 0.3 and 500 Hz, respectively.
2. Dark-adapt the mice overnight (12 h) in a warm, dedicated darkroom before an ERG acquisition (*see* **Note 20**).
3. Turn on the Espion® system according to the manufacturer's instructions, and prepare the mouse according to Subheading 3.3. Place the mouse on the heating pad at 37 °C attached to the system.
4. Place the gold cup electrode in the mouth of the mouse to serve as a reference electrode and a platinum hypodermic needle electrode in the tail as a ground electrode (*see* **Note 21**). Put a small amount of Vidisic® ophthalmic gel on the cornea of the mouse, and place gold wire electrodes on the cornea.
5. Obtain scotopic ERG responses for flash intensities ranging from −3.3 to 1.5 log cd.s.m$^{-2}$ in 0.3-log-unit increments with averaging of at least 30 trials per intensity. Increase the interval between flashes progressively from 10 to 60 s to allow the complete recovery of b-wave amplitude.
6. After completing the scotopic ERG steps, light-adapt the mouse for 15 min in a steady white background (50 cd.m$^{-2}$).
7. Record photopic ERG responses for flash intensities between −0.30 and 1.5 log cd.s.m$^{-2}$ in 0.20-log-unit increments with averaging of 30 responses obtained with a 2 s inter-stimulus interval.
8. Place the mouse under a warm light to recover.

## 4 Notes

1. The cage for the Akita and Akimba mice tends to be wet and dirty easily due to diabetes-associated polyuria. Therefore, the bedding for these mice should be changed more frequently, or house lesser animals in one cage to maintain animal health.

2. Dry the alcohol properly before puncturing the tail as the alcohol may denature or dilute the blood sample, which may lead to inaccurate readings of the plasma glucose level.
3. Apply the pressure for at least 30 s at the snipped area using an alcohol swab to prevent excessive bleeding.
4. Use DNase-free water for the reagent preparation in genotyping to avoid DNA degradation in the samples. Dilution of TaqMan® GTXpress Master Mix and TaqMan® genotyping assay mix is to be done in 1× TE buffer.
5. Observe the purple tracking dye in each well for a uniform filling of the solution.
6. Do not put too many eye drops onto the eye, and make sure the eye drop solution does not flow into the nostril of the animal. The solution may block the animal's airway, leading to respiratory distress.
7. Applying Vidisic® ophthalmic gel as soon as possible after the anesthesia is essential to prevent the formation of anesthesia-induced cataract. The eyes should be kept moist using Vidisic® ophthalmic gel throughout the procedures to avoid developing cataracts.
8. Do not press the objective too hard on the eye. Pressing the objective too hard on the eye can affect the image quality as it changes the cornea shape and increases pressure in the eye.
9. Keeping the anesthetized mouse warm is essential as general anesthesia affects the body's thermoregulation system. Leaving the animals in a cold place can lead to hypothermia and death.
10. The OCT images can be taken in the same session with fundus and FFA imaging.
11. The respiratory movement of the animal can significantly affect the image quality. A proper dose of anesthesia to minimize the movement or immobilize the head of the animal can improve the image quality.
12. The head of the mouse should be immobilized with the head holder to minimize the movement during scanning to get an optimal image quality. Respiratory movement during the scanning can significantly affect the quality.
13. The scanning plane of the retina should be perpendicular to the light path. Tilted or poorly positioned retina gives unevenly focused or blurred angiogram images.
14. The $12 \times 12$ mm$^2$ configuration is referenced to the size of a human eye axial length of ~24 mm. Hence, the actual area of the scanned area in the mouse eye is approximately around $1.6 \times 1.6$ mm$^2$ (axial length of the mouse eye ~3.2 mm).

15. The coverslip should be parallel to the microscope objective (perpendicular to the light path). Tilted coverslip may give reflection or unevenly focused image quality. In addition, the coverslip should be changed frequently as the dirty coverslip will block part of the images or produce artifacts.
16. Do not transfer the blood through the attached needle, as it can destroy the blood cells due to sheer pressure.
17. Precool the centrifuge at 4 °C to avoid temperature change of the sample. A frequent temperature change can reduce the yield of the live cells.
18. The FITC-labelled leukocytes must be kept in a tube wrapped with aluminum foil to prevent photobleaching. Exposure to light may alter the fluorophore of the FITC, resulting in fluorescence fading. The labelled cells should be kept in the dark as much as possible.
19. By warming the mouse, the tail veins are more prominent and prevent them from collapsing. It allows the researchers to see the vessels clearly and makes performing a successful tail vein injection easier.
20. The animals and machines must be prepared under dim red light. Cover all the light sources such as PC monitor, switch indicator lights, and headlamp with a red filter sheet.
21. Extra care should be taken to handle electrode needles in the dark not to injure yourself. Use a proper headlight with a red color light to see and avoid direct hand positioning near the needle.

## References


1. International Diabetes Federation (IDF) (2021) IDF diabetes atlas, 10th edn. IDF, Brussels
2. Cheung AK, Fung MK, Lo AC, Lam TT, So KF, Chung SS, Chung SK (2005) Aldose reductase deficiency prevents diabetes-induced blood-retinal barrier breakdown, apoptosis, and glial reactivation in the retina of db/db mice. Diabetes 54(11):3119–3125. https://doi.org/10.2337/diabetes.54.11.3119
3. Midena E, Segato T, Radin S, di Giorgio G, Meneghini F, Piermarocchi S, Belloni AS (1989) Studies on the retina of the diabetic db/db mouse. I. Endothelial cell-pericyte ratio. Ophthalmic Res 21(2):106–111. https://doi.org/10.1159/000266787
4. Rakoczy EP, Ali Rahman IS, Binz N, Li CR, Vagaja NN, de Pinho M, Lai CM (2010) Characterization of a mouse model of hyperglycemia and retinal neovascularization. Am J Pathol 177(5):2659–2670. https://doi.org/10.2353/ajpath.2010.090883
5. Chaurasia SS, Lim RR, Parikh BH, Wey YS, Tun BB, Wong TY, Luu CD, Agrawal R, Ghosh A, Mortellaro A, Rackoczy E, Mohan RR, Barathi VA (2018) The NLRP3 inflammasome may contribute to pathologic neovascularization in the advanced stages of diabetic retinopathy. Sci Rep 8(1):2847. https://doi.org/10.1038/s41598-018-21198-z
6. Wisniewska-Kruk J, Klaassen I, Vogels IM, Magno AL, Lai CM, Van Noorden CJ, Schlingemann RO, Rakoczy EP (2014) Molecular analysis of blood-retinal barrier loss in the Akimba mouse, a model of advanced diabetic retinopathy. Exp Eye Res 122:123–131. https://doi.org/10.1016/j.exer.2014.03.005
7. Binz N, Rakoczy EP, Ali Rahman IS, Vagaja NN, Lai CM (2016) Biomarkers for diabetic retinopathy – could endothelin 2 be part of the answer? PLoS One 11(8):e0160442. https://doi.org/10.1371/journal.pone.0160442
8. Mendel TA, Clabough EB, Kao DS, Demidova-Rice TN, Durham JT, Zotter BC,

Seaman SA, Cronk SM, Rakoczy EP, Katz AJ, Herman IM, Peirce SM, Yates PA (2013) Pericytes derived from adipose-derived stem cells protect against retinal vasculopathy. PLoS One 8(5):e65691. https://doi.org/10.1371/journal.pone.0065691

9. Van Bergen T, Hu TT, Etienne I, Reyns GE, Moons L, Feyen JHM (2017) Neutralization of placental growth factor as a novel treatment option in diabetic retinopathy. Exp Eye Res 165:136–150. https://doi.org/10.1016/j.exer.2017.09.012
10. Matthews JR, Schlaich MP, Rakoczy EP, Matthews VB, Herat LY (2022) The effect of SGLT2 inhibition on diabetic kidney disease in a model of diabetic retinopathy. Biomedicine 10(3). https://doi.org/10.3390/biomedicines10030522
11. Cronk SM, Kelly-Goss MR, Ray HC, Mendel TA, Hoehn KL, Bruce AC, Dey BK, Guendel AM, Tavakol DN, Herman IM, Peirce SM, Yates PA (2015) Adipose-derived stem cells from diabetic mice show impaired vascular stabilization in a murine model of diabetic retinopathy. Stem Cells Transl Med 4(5): 459–467. https://doi.org/10.5966/sctm.2014-0108
12. Barber AJ, Antonetti DA, Kern TS, Reiter CE, Soans RS, Krady JK, Levison SW, Gardner TW, Bronson SK (2005) The Ins2Akita mouse as a model of early retinal complications in diabetes. Invest Ophthalmol Vis Sci 46(6): 2210–2218. https://doi.org/10.1167/iovs.04-1340
13. Gastinger MJ, Singh RS, Barber AJ (2006) Loss of cholinergic and dopaminergic amacrine cells in streptozotocin-diabetic rat and Ins2Akita-diabetic mouse retinas. Invest Ophthalmol Vis Sci 47(7):3143–3150. https://doi.org/10.1167/iovs.05-1376
14. Van Eeden PE, Tee LB, Lukehurst S, Lai CM, Rakoczy EP, Beazley LD, Dunlop SA (2006) Early vascular and neuronal changes in a VEGF transgenic mouse model of retinal neovascularization. Invest Ophthalmol Vis Sci 47(10): 4638–4645. https://doi.org/10.1167/iovs.06-0251
15. Shen WY, Lai CM, Graham CE, Binz N, Lai YK, Eade J, Guidolin D, Ribatti D, Dunlop SA, Rakoczy PE (2006) Long-term global retinal microvascular changes in a transgenic vascular endothelial growth factor mouse model. Diabetologia 49(7):1690–1701. https://doi.org/10.1007/s00125-006-0274-8
16. Alex AF, Heiduschka P, Eter N (2013) Retinal fundus imaging in mouse models of retinal diseases. Methods Mol Biol 935:41–67. https://doi.org/10.1007/978-1-62703-080-9_3
17. Wigg JP, Zhang H, Yang D (2015) A quantitative and standardized method for the evaluation of choroidal neovascularization using MICRON III fluorescein angiograms in rats. PLoS One 10(5):e0128418. https://doi.org/10.1371/journal.pone.0128418
18. Ochakovski GA, Fischer MD (2019) Phenotyping of mouse models with OCT. Methods Mol Biol 1834:285–291. https://doi.org/10.1007/978-1-4939-8669-9_18
19. Smith CA, Hooper ML, Chauhan BC (2019) Optical coherence tomography angiography in mice: quantitative analysis after experimental models of retinal damage. Invest Ophthalmol Vis Sci 60(5):1556–1565. https://doi.org/10.1167/iovs.18-26441
20. Duggan E, Smith CA, Hooper ML, Chauhan BC (2020) Colocalization of optical coherence tomography angiography with histology in the mouse retina. Microvasc Res 132:104055. https://doi.org/10.1016/j.mvr.2020.104055
21. Liu W, Luisi J, Liu H, Motamedi M, Zhang W (2017) OCT-angiography for non-invasive monitoring of neuronal and vascular structure in mouse retina: implication for characterization of retinal neurovascular coupling. EC Ophthalmol 5(3):89–98
22. Fujii H (1994) Visualisation of retinal blood flow by laser speckle flow-graphy. Med Biol Eng Comput 32(3):302–304. https://doi.org/10.1007/BF02512526
23. Kobayashi T, Shiba T, Okamoto K, Usui T, Hori Y (2022) Characteristics of laterality in the optic nerve head microcirculation obtained by laser speckle flowgraphy in healthy subjects. Graefes Arch Clin Exp Ophthalmol. https://doi.org/10.1007/s00417-022-05631-8
24. Agrawal R, Tun SBB, Balne PK, Zhu HY, Khandelwal N, Barathi VA (2020) Fluorescein labeled leukocytes for in vivo imaging of retinal vascular inflammation and infiltrating leukocytes in laser-induced choroidal neovascularization model. Ocul Immunol Inflamm 28(1): 7–13. https://doi.org/10.1080/09273948.2018.1429637
25. Agrawal R, Balne PK, Tun SBB, Sia Wey Y, Khandelwal N, Barathi VA (2017) Fluorescent dye labeling of erythrocytes and leukocytes for studying the flow dynamics in mouse retinal circulation. J Vis Exp 125. https://doi.org/10.3791/55495
26. McCulloch DL, Marmor MF, Brigell MG, Hamilton R, Holder GE, Tzekov R, Bach M (2015) ISCEV standard for full-field clinical electroretinography (2015 update). Doc Ophthalmol 130(1):1–12. https://doi.org/10.1007/s10633-014-9473-7

# Chapter 3

# Oxygen-Induced Retinopathy in the Rat: An Animal Model to Study the Proliferative Retinal Vascular Pathology


## Yu-Fan Chuang and Fan-Li Lin



## Abstract

Diabetic retinopathy (DR) is one of the leading causes of vision loss worldwide. There are numerous animal models available for developing new ocular therapeutics and drug screening and to investigate the pathological processes involved in DR. Among those animal models, the oxygen-induced retinopathy (OIR) model, though originally developed as a model for retinopathy of prematurity, has also been used to investigate angiogenesis in proliferative DR with the phenomenon of ischemic avascular zones and pre-retinal neovascularization it demonstrated. Briefly, neonatal rodents are exposed to hyperoxia to induce vaso-obliteration. Upon removal from hyperoxia, hypoxia develops in the retina that eventually results in neovascularization. The OIR model is mostly used in small rodents such as mice and rats. Here, we describe a detailed experimental protocol of rat OIR model and the subsequent assessment of abnormal vasculature. By illustrating the vasculoprotective and anti-angiogenic activities of the treatment, OIR model might advance to a new platform for investigating novel ocular therapeutic strategies for DR.



**Key words** Diabetic retinopathy, Proliferative diabetic retinopathy, Retinal neovascularization, Oxygen-induced retinopathy, Hyperoxia, Hypoxia, Vaso-obliteration


## 1 Introduction

It is estimated by the International Diabetes Federation that by 2045, the number of people with diabetes mellitus (DM) will be 700 million [1]. Among the microvascular complications of DM, diabetic retinopathy (DR) is the most common one [2]. Diabetic retinopathy (DR) is one of the leading causes of blindness worldwide [3, 4]. It has been reported that hyperglycemia plays a crucial role in the progression of DR [5, 6]. DR is mainly classified into two stages depending on the presence of neovascularization (NV) including proliferative DR (PDR), with neovascularization, and non-proliferative DR (NPDR), absence of neovascularization [4, 7, 8]. Therefore, the aberrant NV observed in the retina is the main hallmark of PDR which distinguishes it from NPDR. Chronic hyperglycemia triggers the progression of DR into PDR


Guei-Sheung Liu and Jiang-Hui Wang (eds.), *Diabetic Retinopathy: Methods and Protocols*, Methods in Molecular Biology, vol. 2678, https://doi.org/10.1007/978-1-0716-3255-0_3,

by promoting vascular damage which leads to retinal ischemia that eventually induces NV. Rodent oxygen-induced retinopathy (OIR) model is a useful system that enables the simulation of ischemic-induced NV and is applicable to studying retinal diseases with vascular changes including DR [9–11]. There are two phases of OIR models: during the first phase, hyperoxic condition induces vascular obliteration, and the second phase is initiated by hypoxic condition when removing animals from the hyperoxia to normoxia. During the second phase, hypoxia enhances the expression of angiogenic factors such as VEGF that promotes angiogenesis in avascular areas in the retina that mimics the vascular changes in DR [9, 12, 13]. The mostly used animals for OIR models are rodents including mouse and rat, while the induction mode is different among them. Chronic hyperoxia is used in the mouse model in which the neonatal mice (7-day-old, P7) are exposed to 75% oxygen from P7 to P12 (5 days) and then return them to room air at P12. During the time in room air (P12–P17), neovascularization initiates at P14 and peaks at P17. The complete resolution in which the aberrant vessels are replaced by normal vasculature is achieved at P25 [14]. Different from the mouse OIR model, Penn et al. successfully demonstrated a rat OIR model using alternating hyperoxia/hypoxia approach [10]. In the rat OIR model, neonatal rats are exposed to 50–80% oxygen at birth for 24 h, followed by 10–40% oxygen for 24 h, and this sustained hyperoxia/hypoxia cycle will continue for 14 days. Exposure to high oxygen during the developmental period suppresses the normal vasculature formation, resulting in avascular regions (vaso-obliteration) within the peripheral retina. Then, the rats are returned to room air on P14 and sacrificed on P18, the time point of maximum retinal neovascularization in this model, for abnormal vasculature validation [10, 15]. Here, we demonstrated a detailed rat OIR model in which neonatal rats were exposed to 80% oxygen at birth (within 12 h of birth, P0) for 21 h and then exposed to room air (21% oxygen) for 3 h, and this daily cycle continued until P14. Rats were removed to room air from P14 to P18 (4 days) and sacrificed on P18. Retinal NV and vaso-obliteration were subsequently validated via retina dissection and retinal flat-mount staining at P18. We have coupled this rat OIR model with gene therapy to suggest an innovative promising approach for the treatment of retinal neovascularization [16, 17].

## 2 Materials

### *2.1 Oxygen-Induced Retinopathy Rat Model*

1. Sprague Dawley rat neonates.
2. Nursing Sprague Dawley rat mothers.

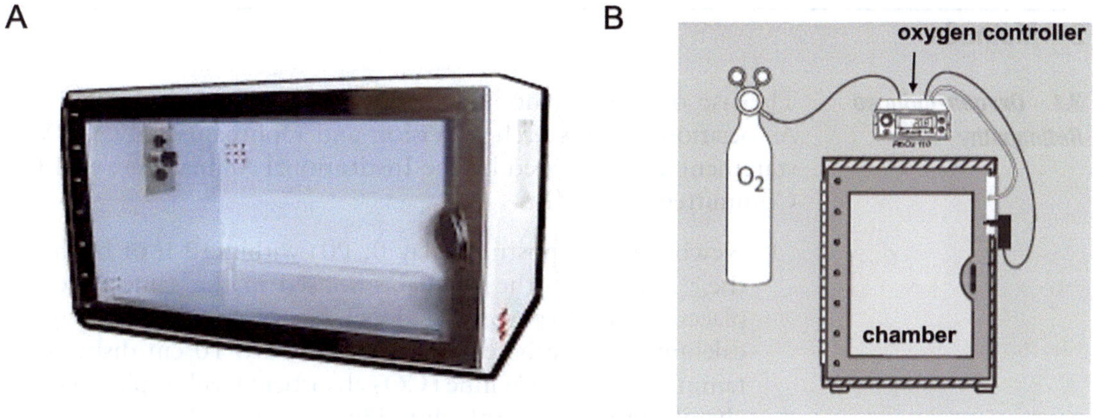


**Fig. 1** Illustration of hyperoxic chamber and oxygen controller system used in OIR model. (**a**) The interior dimensions of hyperoxic chamber were as follows: length 168 cm, width 51 cm, and height 51 cm. (**b**) A schematic illustration of oxygen controller system. The oxygen level inside the chamber can be controlled by an oxygen controller with air input from oxygen cylinder

3. Oxygen cylinder.
4. Oxygen controller.
5. Hyperoxic chamber. The hyperoxic chamber was designed to accommodate two rat cages at the same time; the interior dimensions are length 66″ (168 cm), width 20″ (51 cm), and height 20″ (51 cm) (Fig. 1a). The atmosphere can be controlled by an oxygen controller to reach expected oxygen level with air input from oxygen cylinder (Fig. 1b).
6. Calcium chloride-based desiccant.
7. Medical soda lime.

### 2.2 Immunostaining of the Retinal Vasculature

Prepare all solutions using ultrapure water and analytical grade reagents.

1. Vannas scissor: Straight, blade 7 mm, 8 cm.
2. High-precision tweezers: Straight, very fine, sharp, superior finish, overall length 11 cm.
3. 1× phosphate-buffered saline (PBS): 137 mM NaCl, 2.7 mM KCl, 10 mM $Na_2HPO_4$, 1.8 mM $KH_2PO_4$, pH 7.4.
4. Ethanol: 70% solution in water.
5. Triton-X: 1% solution in water.
6. Isolectin GS-$IB_4$ from *Griffonia simplicifolia*, Alexa Fluor™ 488 conjugate (isolectin B4-Alexa488): 1% solution in PBS.
7. Paraformaldehyde (PFA): 4% in PBS, pH 7.
8. Fluorescent mounting medium.

## 3 Methods

### 3.1 Oxygen-Induced Retinopathy

The use of animals should be conducted in accordance with the Association for Research in Vision and Ophthalmology (ARVO) statement and approved by the Institutional Animal Care and Use Committee (IACUC).

1. Newborn rats (postnatal day 0; P0) within 12 h of birth and their nursing mothers will be housed in the same cage and placed in the hyperoxic chamber (*see* **Note 1**). Four bags of calcium chloride-based desiccant and two 10 cm dishes containing medical soda lime ($CO_2$ absorbent) will be placed in the chamber to maintain the humidity (aim for 70–75% humidity) and to prevent $CO_2$ retention (*see* **Note 2**).
2. Set the oxygen level of the oxygen controller to 80% and the output of oxygen cylinder to 18 psi. It usually takes around 10 min to reach the expected oxygen level. P0 pups and nursing mother are exposed to 80% oxygen for 21 h followed by 21% oxygen (room air) for 3 h (P1). This daily cycle continues for 14 days (P1–P14) (Fig. 2), which means the pups and mother are exposed to room air (chamber open) for 3 h each day from P1 to P13 (*see* **Note 3**).
3. Cage change should be done at least 5–6 days after rat birth, which will be repeated every 3 days, as well as the updating of the desiccant and medical soda lime (*see* **Note 4**).
4. Animals were returned to room air at P14 and sacrificed to harvest retina at P18 (*see* **Note 5**).

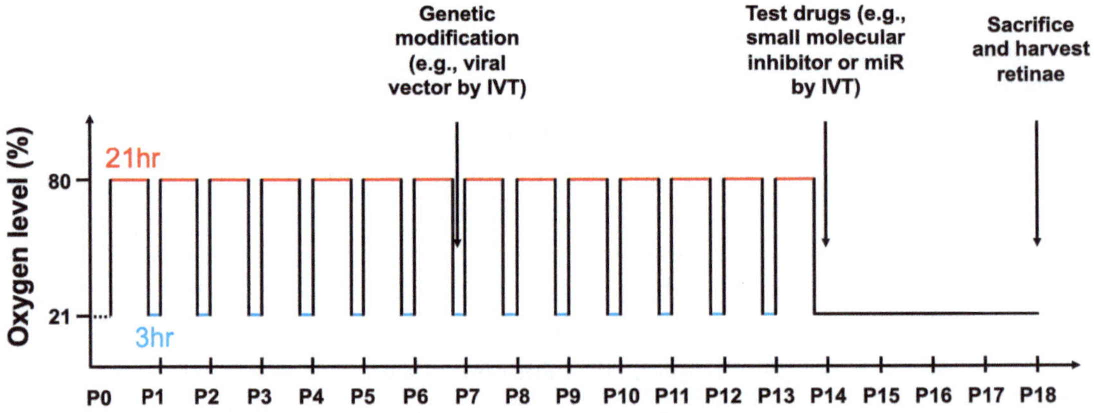


**Fig. 2** Experimental protocol for OIR rat model. Sprague Dawley litters (within 12 h of birth; P0) and their nursing mother were exposed to daily cycles of 80% oxygen for 21 h and room air (21% oxygen) for 3 h in hyperoxic chamber until P14. At P14, animals were returned to room air until P18, when they were sacrificed to harvest retina. Intervention with genetic modification or drug treatment can be intravitreally injected at P7 or P14, respectively

### 3.2 Dissection of Retina

Ensure the scissor is sharp. Vannas scissor and high-precision tweezers are recommended for small tissue dissection including the eyeball and retina.

1. Place humanely sacrificed rat pup on its side, and enucleate the eye by scissors and forceps. Place scissors beneath the enucleated eye, cut the optic nerve, and lift the eye out of the socket.
2. Make a puncture at the cornea using a 29G needle, and transfer the eye to a 1.5 mL tube containing 4% PFA for 1 h at 4 °C.
3. Discard the PFA and re-fill with 500 μL cold 1× PBS; store the tube at 4 °C.
4. Transfer the eyes into a dry Petri dish on ice using a plastic Pasteur pipette that has been cut to widen the bore, and place it under a dissecting microscope.
5. Remove fat and muscle tissue around the eye using scissor and tweezers.
6. Cut along the line slightly below the limbus to remove the cornea and iris (*see* **Note 6**).
7. Insert the tweezer between the sclera and retina and hold the sclera. Gently tear and remove the sclera by another tweezer (*see* **Note** 7).
8. Remove the lens and clean the vitreous using forceps (*see* **Note 8**).
9. Cup-shaped retina was radially cut into a petal shape with four to five incisions approximately 2/3 of the radius of the retina.
10. Transfer to a 1.5 mL tube containing 300 μL cold 1× PBS by a wide-bore plastic Pasteur pipette.

### 3.3 Isolectin B4-Alexa488 Staining of the Retinal Vasculature

1. Discard the PBS and incubate the retina with 300 μL 70% ethanol for 30 min at 4 °C (at this stage, the retina is still in the same tube from **step** 7).
2. Wash with 300 μL 1× PBS once by gently pipetting, then remove PBS, and incubate the retina with 300 μL 1% Triton-X for 20 min at room temperature.
3. Wash with 300 μL 1× PBS once by gently pipetting.
4. Discard the PBS and incubate the retina with 100 μL 1% isolectin B4-Alexa488 antibody on a rocker overnight at 4 °C (*see* **Note 9**).
5. Discard the antibody and wash the retina with 500 μL 1× PBS four times, 7 min each time on a rocker at room temperature.
6. Transfer the retina onto glass slide using a wide-bore plastic Pasteur pipette, flatten the retina, and then remove excess PBS with absorbent paper.

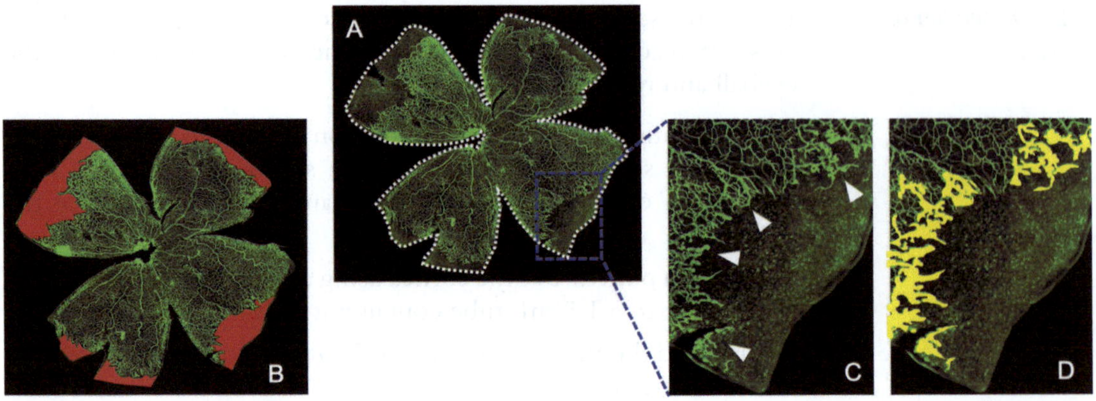


**Fig. 3** Identification of vaso-obliteration and neovascularization area in OIR rat retina. (**a**) Isolectin B4 whole mount staining revealed the retinal vasculature in P18 OIR-conditioned retina. The whole retina area was labelled with white dotted line. (**b**) Vaso-obliteration was simultaneously evaluated using the same sample presented in (**a**). The red areas represent the vaso-obliteration (avascular) area in the retina. (**c**) The inset shows selected area (blue dotted box) from (**a**) at high magnification. Arrowheads indicated neovascular tufts (abnormal neovascularization) that occurred on the margin of the developing blood vessels on the surface of the retina. (**d**) The yellow areas represent the neovascularization area from (**c**)

7. Mount the retina with 50 μL fluorescent mounting medium, and gently place the cover glass over the retina, ensuring the retina is flat without bubbles.
8. Keep slides from light and store slides overnight at 4 °C to enable the mounting medium to set.
9. Image retina using a fluorescence microscope or confocal microscope, and identify the vaso-obliteration (Fig. 3a, b) and retinal neovascularization area (Fig. 3c, d) (*see* **Note 10**).

## 4 Notes

1. The usual litter size is 10–14 pups. It is recommended not to leave more than 15 pups ($\leq 14$) for each experimental batch to ensure the pups can grow normally so as retinal vasculature development by sufficient breastfeeding. The litter should not be disturbed for a few days after birth; if transfer of P0 pups into a new cage is essential, follow the steps below: (1) Transfer "dirty" enrichment across (familiar smells on tunnel, wood chew, etc.). (2) Transfer the mother across. (3) Cup your hands underneath all of the nesting material (wood wool and paper towel) that is holding the pups, and transfer it to the same position in the clean cage, all without touching the pups unless necessary. (4) Gather up any nesting material left behind in the dirty cage, and place it on top of the litter in the clean cage. Leave some animal diet at the bottom of a new cage in case the nursing rat is not able to breastfeed.

**Table 1**
**Humidity and temperature monitoring sheet**

| Chamber to be opened 3 h per day from P1 to P13 | | | | | | | |
|---|---|---|---|---|---|---|---|
| **Date** | **Humidity** | **Temp** | **Animal age** | **Chamber open** | | **Chamber close** | |
| – | – | – | – | Time | Technician | Time | Technician |
| | | | P0 | Into chamber | | | |
| | | | P1 | | | | |
| | | | P2 | | | | |
| | | | P3 | | | | |
| | | | P4 | | | | |
| | | | P5 | | | | |
| | | | P6 | | | | |
| | | | P7 | | | | |
| | | | P8 | | | | |
| | | | P9 | | | | |
| | | | P10 | | | | |
| | | | P11 | | | | |
| | | | P12 | | | | |
| | | | P13 | | | | |
| | | | P14 | | | Into normal housing | |
| P15: normal housing | | | | | | | |
| P16: normal housing | | | | | | | |
| P17: normal housing | | | | | | | |
| P18: normal housing and use for exp | | | | | | | |

Notes:
1. To turn off the oxygen, do not touch the cylinder; only touch the regulator on top of the chamber
2. Acceptable oxygen saturation range of 75–85%. If outside of the above parameter, contact XXX
3. Output of O2 set as 18 psi
4. Make sure water/food levels and leave some food at the bottom of the cage

2. To prevent $CO_2$ retention, the quantity of medical soda lime can be loaded as much as possible in a 10 cm dish (lid off). Color change can be from white to violet/purple. Recording of humidity and temperature is essential; *see* Table 1.
3. Acceptable oxygen saturation range is 75–85%. Recording open/close time of chamber is optional; *see* Table 1. Observation of possible side effects for both nursing mother and pups is required, the parameters including activity (reduced activity, agitated, or aggressive), hydration, and oxygen toxicity

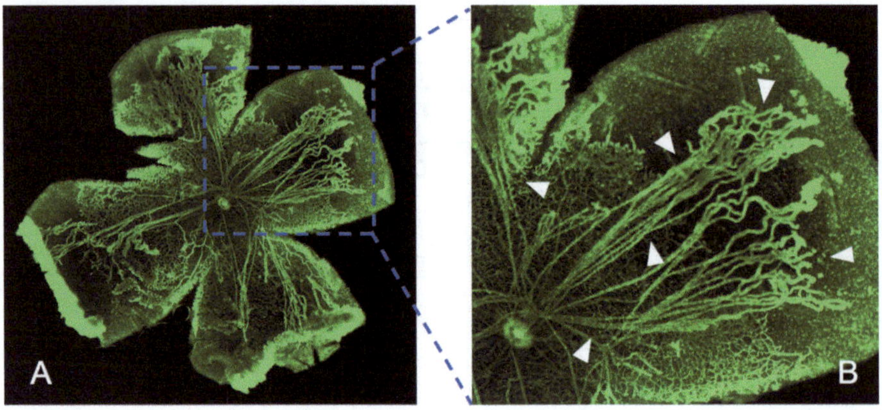


**Fig. 4** Hyaloid vessels that existed in vitreous area in OIR rat retina. (**a**) Retinal whole mount stained with isolectin B4-Alexa488 revealed the retinal vasculature P18 OIR rat retina, and (**b**) inset showed the selected area at high magnification. Arrowheads indicated hyaloid vessels distributed in the unremoved vitreous

(abnormal breath and neurological symptoms such as twitching or tremors). Visual check for pups without touching the pups unless necessary.

4. In principle, the litter should not be disturbed for at least 7 days after birth. However, we notice that humidity is usually reaching ~75% at P5 or P6, so a cage change was suggested at this time point that helps to lower the humidity. We also found that humidity is difficult to control (> 75%) after P5 or P6 till P14, so a frequent cage change (every 3 days) with the update of desiccant (four bags) and medical soda lime (two 10 cm dishes) is recommended.
5. If a drug test for anti-angiogenic activity is required, the compound such as small molecular inhibitor or miRNA can be intravitreally injected at P14. If genetic modification such as viral-based gene therapy is considered, the materials can be intravitreally injected at P7 (the eyelid of neonatal needs to be opened using a scalpel blade) (Fig. 2). Beware; the surgery should be done under the surveillance of a veterinary.
6. The limbus is the juncture of the cornea and the sclera. The inner zone of this junction contains the ciliary body and iris. The ciliary body is the anterior continuation of the retina and choroid and is attached to the sclera. Therefore, a circle cut around the limbus helps to dissociate the retina.
7. Cutting where optic nerve passes through the sclera using the Vannas scissor will help to release the retina.
8. Remove the vitreous as much as possible by blunted-edge forceps, since the remaining vitreous may hinder antibody access to antigens during immunostaining. Moreover, the hyaloid vessels, a transient network of blood vessels that existed in the vitreous, can also affect the result of isolectin B4-stained retinal vasculature (Fig. 4).

9. A secondary antibody is not required when immunostaining with isolectin B4-Alexa488, a directly conjugated primary antibody. Thus, retinas can be mounted directly (**step** 7). If co-immunostaining is considered, another primary antibody can be applied in **step 5** and incubated on a rocker overnight at 4 °C. The next day, repeat wash as **step 5**, and incubate with appropriate fluorescent secondary antibody (diluted in PBS, generally 1/200) for 2 h at room temperature. Then repeat **steps 5–9** for slide preparation.
10. The neovascular tufts that occurred on the margin of the developing blood vessels on the surface of the retina indicate abnormal neovascularization. The differences of vaso-obliteration and neovascularization area between vehicle- or drug-treated groups illustrate the vasculoprotective and anti-angiogenic activities of the drug, respectively.

## References


1. Saeedi P, Petersohn I, Salpea P, Malanda B, Karuranga S, Unwin N, Colagiuri S, Guariguata L, Motala AA, Ogurtsova K, Shaw JE, Bright D, Williams R, Committee IDFDA (2019) Global and regional diabetes prevalence estimates for 2019 and projections for 2030 and 2045: results from the International Diabetes Federation Diabetes Atlas, 9(th) edition. Diabetes Res Clin Pract 157:107843. https://doi.org/10.1016/j.diabres.2019.107843
2. Khalil H (2017) Diabetes microvascular complications – a clinical update. Diabetes Metab Syndr 11(Suppl 1):S133–S139. https://doi.org/10.1016/j.dsx.2016.12.022
3. Leasher JL, Bourne RR, Flaxman SR, Jonas JB, Keeffe J, Naidoo K, Pesudovs K, Price H, White RA, Wong TY, Resnikoff S, Taylor HR, Vision Loss Expert Group of the Global Burden of Disease Study (2016) Global estimates on the number of people blind or visually impaired by diabetic retinopathy: a meta-analysis from 1990 to 2010. Diabetes Care 39(9):1643–1649. https://doi.org/10.2337/dc15-2171
4. Cheung N, Mitchell P, Wong TY (2010) Diabetic retinopathy. Lancet 376(9735):124–136. https://doi.org/10.1016/S0140-6736(09)62124-3
5. Cunha-Vaz J (2007) Characterization and relevance of different diabetic retinopathy phenotypes. Dev Ophthalmol 39:13–30. https://doi.org/10.1159/000098497
6. Duh EJ, Sun JK, Stitt AW (2017) Diabetic retinopathy: current understanding, mechanisms, and treatment strategies. JCI Insight 2(14). https://doi.org/10.1172/jci.insight.93751
7. Liu CH, Wang Z, Sun Y, Chen J (2017) Animal models of ocular angiogenesis: from development to pathologies. FASEB J 31(11):4665–4681. https://doi.org/10.1096/fj.201700336R
8. Robinson R, Barathi VA, Chaurasia SS, Wong TY, Kern TS (2012) Update on animal models of diabetic retinopathy: from molecular approaches to mice and higher mammals. Dis Model Mech 5(4):444–456. https://doi.org/10.1242/dmm.009597
9. Scott A, Fruttiger M (2010) Oxygen-induced retinopathy: a model for vascular pathology in the retina. Eye (Lond) 24(3):416–421. https://doi.org/10.1038/eye.2009.306
10. Penn JS, Henry MM, Tolman BL (1994) Exposure to alternating hypoxia and hyperoxia causes severe proliferative retinopathy in the newborn rat. Pediatr Res 36(6):724–731. https://doi.org/10.1203/00006450-199412000-00007
11. Penn JS, Tolman BL, Lowery LA (1993) Variable oxygen exposure causes preretinal neovascularization in the newborn rat. Invest Ophthalmol Vis Sci 34(3):576–585
12. Lange C, Ehlken C, Stahl A, Martin G, Hansen L, Agostini HT (2009) Kinetics of retinal vaso-obliteration and neovascularisation in the oxygen-induced retinopathy (OIR) mouse model. Graefes Arch Clin Exp Ophthalmol 247(9):1205–1211. https://doi.org/10.1007/s00417-009-1116-4
13. Kern TS (2007) Contributions of inflammatory processes to the development of the early stages of diabetic retinopathy. Exp Diabetes Res 2007:95103. https://doi.org/10.1155/2007/95103

14. Smith LE, Wesolowski E, McLellan A, Kostyk SK, D'Amato R, Sullivan R, D'Amore PA (1994) Oxygen-induced retinopathy in the mouse. Invest Ophthalmol Vis Sci 35(1): 101–111
15. Hartnett ME, Martiniuk D, Byfield G, Geisen P, Zeng G, Bautch VL (2008) Neutralizing VEGF decreases tortuosity and alters endothelial cell division orientation in arterioles and veins in a rat model of ROP: relevance to plus disease. Invest Ophthalmol Vis Sci 49(7):3107–3114. https://doi.org/10.1167/iovs.08-1780
16. Chen J, Lin FL, Leung JYK, Tu L, Wang JH, Chuang YF, Li F, Shen HH, Dusting GJ, Wong VHY, Lisowski L, Hewitt AW, Bui BV, Zhong J, Liu GS (2021) A drug-tunable Flt23k gene therapy for controlled intervention in retinal neovascularization. Angiogenesis 24(1):97–110. https://doi.org/10.1007/s10456-020-09745-7
17. Wang JH, Chuang YF, Chen J, Singh V, Lin FL, Wilson R, Tu L, Ma C, Wong RCB, Wang PY, Zhong J, Hewitt AW, van Wijngaarden P, Dusting GJ, Liu GS (2022) An integrative multi-omics analysis reveals MicroRNA-143 as potential therapeutics to attenuate retinal angiogenesis. Nucleic Acid Ther. https://doi.org/10.1089/nat.2021.0111

# Chapter 4

# Retinal Assessment Using In Vivo Electroretinography and Optical Coherence Tomography in Rodent Models of Diabetes

**Da Zhao, Pei Ying Lee, Vickie H. Y. Wong, Tomoharu Nishimura, Anh Hoang, Katie K. N. Tran, Anna K. van Koeverden, Brianna C. Afiat, Christine T. O. Nguyen, and Bang V. Bui**

## Abstract

Electroretinography and optical coherence tomography imaging allow for non-invasive quantitative assessment of the retina. These approaches have become mainstays for identifying the very earliest impact of hyperglycemia on retinal function and structure in animal models of diabetic eye disease. Moreover, they are essential for assessing the safety and efficacy of novel treatment approaches for diabetic retinopathy. Here, we describe approaches for in vivo electroretinography and optical coherence tomography imaging in rodent models of diabetes.

**Key words** Small animal models, Rodents, In vivo retinal assessment, Electroretinogram, Retinal function, Retinal structure, Optical coherence tomography

## 1 Introduction

A range of in vivo tools have been developed to assess retinal changes in models of diabetic retinopathy (DR), in particular in rats and mice. Increasingly, studies of retinal disease-causing genes or potential new therapies using in vivo models demand a functional readout of retinal or visual pathway health. Among non-invasive behavioral [1] and electrophysiology approaches [2], perhaps the most common in vivo tool that allows for the longitudinal functional quantification of disease or treatment progress is the electroretinogram (ERG) [3]. More recently, the ease with which gross retinal structure and blood vessels can be imaged using optical coherence tomography (OCT) and OCT angiography (OCTA) has seen the use of this technology adopted in diabetic eye research, to provide information complementary to functional measures.

Guei-Sheung Liu and Jiang-Hui Wang (eds.), *Diabetic Retinopathy: Methods and Protocols*, Methods in Molecular Biology, vol. 2678, https://doi.org/10.1007/978-1-0716-3255-0_4,

The ERG measures the electrical responses of the retina to light stimulation and provides a means to simultaneously quantify the functional integrity of key retinal cell classes. A coordinated series of ionic fluxes initiated by light onset and offset produce detectable changes in voltage that can be measured using surface electrodes placed on or near the eye. The resultant waveform is the combination of a series of well-defined components which can be broken down to provide information on light-capturing photoreceptors, bipolar cell interneurons, inhibitory pathways, and inner retinal ganglion cells [4–6].

Spectral domain optical coherence tomography (SD-OCT) returns high-resolution in vivo retinal cross-sections and is becoming common for assessing retinal morphology in rodent models of diabetes [7]. In addition to valuable insights into retinal structure provided by OCT, the advent of OCTA has opened the possibility of quantifying disease or treatment-induced changes in the retinal microvasculature [8].

This manuscript describes a protocol for the recording and analysis of the ERG as well as the imaging of retinal structure and blood vessels using OCT/OCTA, respectively.

## 2 Materials

### *2.1 Electro-retinography*

1. Electrode silver wire (0.3-mm-diameter 99.9% silver round wire).
2. Insulating tape.
3. Chloriding battery (9 V battery).
4. Platinum electrode leads.
5. Electrode holders.
6. Ground electrode.
7. Headlight for animal preparation (Energizer HD+ LED headlight).
8. Mydriatic agent (0.5–1% tropicamide, Alcon).
9. Corneal anesthesia (0.5% proxymetacaine, Alcon).
10. General anesthesia (1.6 mL ketamine [100 mg/mL], 1.6 mL xylazine [20 mg/mL], and 6.8 mL sodium chloride [0.9% w/v]).
11. Contact gel (carmellose sodium, Celluvisc 1.0%, Allergan).
12. Light stimulus via Ganzfeld integrating bowl (36 cm diameter, 13 cm aperture size).
13. Bioamplifier (Grass P511).

14. Data acquisition system (PowerLab ML785 data acquisition system).
15. Data acquisition software (Scope software, version 3.7.6).
16. Faraday cage (2 × 1.2 × 1.2 m).

### 2.2 Optical Coherence Tomography (OCT)/ Optical Coherence Tomography A (OCT-A)

1. Spectral domain optical coherence tomography (SD-OCT, Spectralis OCT2).
2. 25D lens for use on small eyes (mouse) with SD-OCT.
3. Small animal platform.
4. Computer (Windows based) for recording signals and offline analysis.
5. Ocular lubricant.
6. Hydrating gel.
7. OCT analysis software.
8. Microvessel analysis software.

## 3 Methods

### 3.1 Electroretinography Assessment of Scotopic Retinal Function

The following describes the use of dark-adapted full-field ERG in order to examine retinal function as described previously [9, 10].

1. Prepare four electrodes (two active, two reference) for recording from both eyes (Fig. 1a). Prepare the active electrodes by attaching (twisting) silver wire to platinum leads (with 2 cm of the plastic sleeve removed). Fold the silver back so that the tip is rounded. Use masking tape to insulate the connections leaving a 1 mm rounded tip of silver wire exposed to contact the cornea. The two reference electrodes are made the same way, but instead of a 1 mm rounded tip, create a ~4-mm-diameter circle for mice or ~8-mm-diameter circle for rats, which should slide around the equator of the eye without squeezing the eye. Insulate the connection leaving the loop exposed.
2. Chloride the exposed silver tips by immersing them into 0.9% saline bath and connecting the leads to a 9 V battery and completing the circuit using another length of conductive wire. Chloride for 10 s, after which time the silver wire should appear coated with a layer of white salt. Prior to use, lightly dust off any excess salt which can damage the cornea using a lint-free tissue.
3. Connect the silver end of the active electrode to micromanipulators. Connect all electrodes (including the ground needle electrode) to preamplifier input cables.

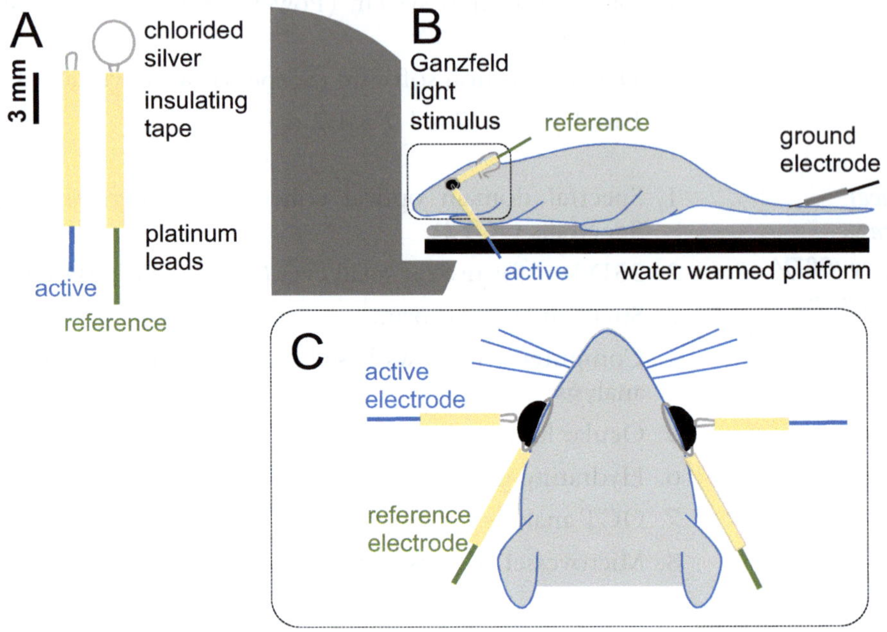


**Fig. 1** Preparation of electroretinogram electrodes and position on the mouse eye. (**a**) Active and reference electrodes are made by connecting pure silver wire to platinum leads and chloriding the silver tips using a 9 V battery for 10 s. Connections are insulated leaving the tips exposed to contact the eye. Fore mice reference electrodes (~3 mm) should be smaller than for rats (~8 mm). (**b**) The rodent is placed on a warmed platform with its eyes positioned in the plane of the Ganzfeld stimulator, with electrodes placed on the eyes and a ground electrode in the tail. (**c**) Recordings are made simultaneously from both eyes, with the active electrodes gently touching the corneal apex and reference electrodes encircling the eye around the equator

4. Dark-adapt rodents overnight prior to recording [9, 10].
5. Minimize exposure to light during anesthesia and animal preparation to ensure maximal retinal sensitivity and to produce high-quality recordings, particularly for small waveforms such as the ganglion cell response. Throughout preparation, keep the room lights off, and use a headlight to provide dim red illumination (17.4 cd.m$^{-2}$, $\lambda$max = 600 nm). This helps to maintain dark adaptation prior to signal recording.
6. Rodents are weighed and anesthetized with a ketamine and xylazine cocktail delivered via an intraperitoneal injection using a 30G needle (*see* **Note 1**).
7. Dilate pupils using one drop of 0.5–1% tropicamide, and place a drop of 0.5% proxymetacaine onto each eye for corneal anesthesia. Leave for 5 s, and dab off any excess fluid from the margin of the eyelids using a lint-free tissue.
8. Place rodent on a platform warmed (37 °C) by using circulating heated water (Lauda). Electrical heating pads will generate electrical noise.

9. Insert the needle electrode into the tail (~2 mm subcutaneously) approximately halfway along the tail (Fig. 1b).
10. Gently place the reference electrode loop around each eye (Fig. 1c), using masking tape or Velcro to secure in place firmly (without affecting respiration) at the nape of the neck. Ensure that the loops don't squeeze the eyes as this will increase eye pressure, which can cause a reduction in signal size.
11. Use micromanipulators to allow for fine adjustment to position the tips of the active electrodes to gently touch the corneal apex (Fig. 1c). Place a small drop of contact gel onto the active electrode tip to improve electrical contact and maintain corneal hydration. Avoid allowing the lubricant to contact both active and reference electrodes. If needed, twist a Kimwipe into a fine tip, and gently wipe along the margin of the lower lid to remove any excess contact gel.
12. Move the stimulus so that the eyes are in the plane of the Ganzfeld sphere. Close the Faraday cage.
13. Allow the animals to rest for at least 5 min in full darkness to recover for full scotopic sensitivity after exposure to the red headlight during animal setup.
13. Stimuli are delivered starting with the dimmest light level ($-5.53$ log cd·s/m$^2$) and progress to the brightest (2.07 log cd·s/m$^2$). Signal averaging may be needed to remove random noise arising from animal-breathing and other small involuntary movements and environment-related sources of noise. Average enough signals for a signal-to-noise ratio of $>5$. Suggested light levels, signal averaging, and interstimulus intervals can be found in Table 1 (*see* **Note 2**).
14. Use a data acquisition system (ML785 PowerLab 8SP) and appropriate software (Scope™) to record signals of 2560 points at a 4 kHz sampling rate with band-pass filter settings of 0.3–1000 Hz ($-3$ dB).
15. For an ERG waveform elicited by a bright flash of light, the very first electronegative component, known as the a-wave, comes from photoreceptors and can be described using a delayed Gaussian function (Fig. 2a inset). This equation models the underlying P3 generated by photoreceptors [11] to return a maximal amplitude (RmP3, μV) indicating the number of photoreceptors (outer segment length or density of non-specific cationic channels). The slope parameter (sensitivity, S, m$^2$.cd$^{-1}$.s$^{-3}$) provides a measure of the efficiency of phototransduction or the rate of amplification.
16. The second and largest component of the ERG waveform (the b-wave) is the envelope of the negative photoreceptor P3 and positive P2, which is dominated by ON bipolar cells. Subtracting the delayed Gaussian P3 model (Fig. 2a blue traces) from

**Table 1**
**Electroretinogram (ERG) light stimulus protocol**

| Light level (log cd.s/m$^2$) | Number of repeats | Time between flashes (s) | ERG component elicited |
|---|---|---|---|
| −5.53 | 20 | 2 | STR |
| −5.31 | 20 | 2 | STR |
| −5.01 | 20 | 2 | STR |
| −4.90 | 20 | 2 | STR |
| −4.31 | 5 | 3 | STR + b-wave |
| −3.36 | 5 | 3 | Rod b-wave |
| −2.75 | 1 | 15 | Rod b-wave, OPs |
| −1.61 | 1 | 30 | Rod b-wave, OPs |
| −0.81 | 1 | 45 | Mixed b-wave, OPs, a-wave |
| 0.72 | 1 | 60 | Mixed b-wave, OPs, a-wave |
| 1.55 | 1 | 120 | Mixed b-wave, OPs, a-wave |
| 2.07 | 2 | 180 | **1st flash:** mixed b-wave, OPs, a-wave<br>**2nd flash:** cone b-wave, OPs |

Different light levels elicit various components of the ERG including the ganglion cell scotopic threshold response (STR), the bipolar cell b-wave, the photoreceptor a-wave, and inner retinal oscillatory potentials (OPs). Repeating bright flashes with very short interstimulus intervals elicits a subsequent response that does not contain contributions from the rod pathway

the raw ERG waveform (Fig. 2a gray traces) reveals the P2 along with inner retinal oscillatory potentials (OPs). Low-pass filtering (50 Hz) will remove the OPs and high-frequency noise leaving the P2 (Fig. 2b gray traces). The amplitude of each P2 waveform (Fig. 2b blue dots) can be described as a function of intensity with a classical log agonist-receptor characteristic [12]. This model (Fig. 2e) returns a maximum amplitude (Vmax), which provides an index of bipolar cell integrity along with a sensitivity parameter (1/K).

17. The OPs, which provide a measure of inner retinal inhibitory circuits involving amacrine cells [13], can be isolated using a band-pass filter (Fig. 2c, 50–180 Hz), after which peak amplitude and implicit time can be extracted.
18. The positive scotopic threshold response (pSTR), which can only be measured at very dim light levels, is a good indicator of retinal ganglion cell function in rodents (Fig. 2d) [9, 14]. Quantify the pSTR by taking its peak amplitude and implicit time.

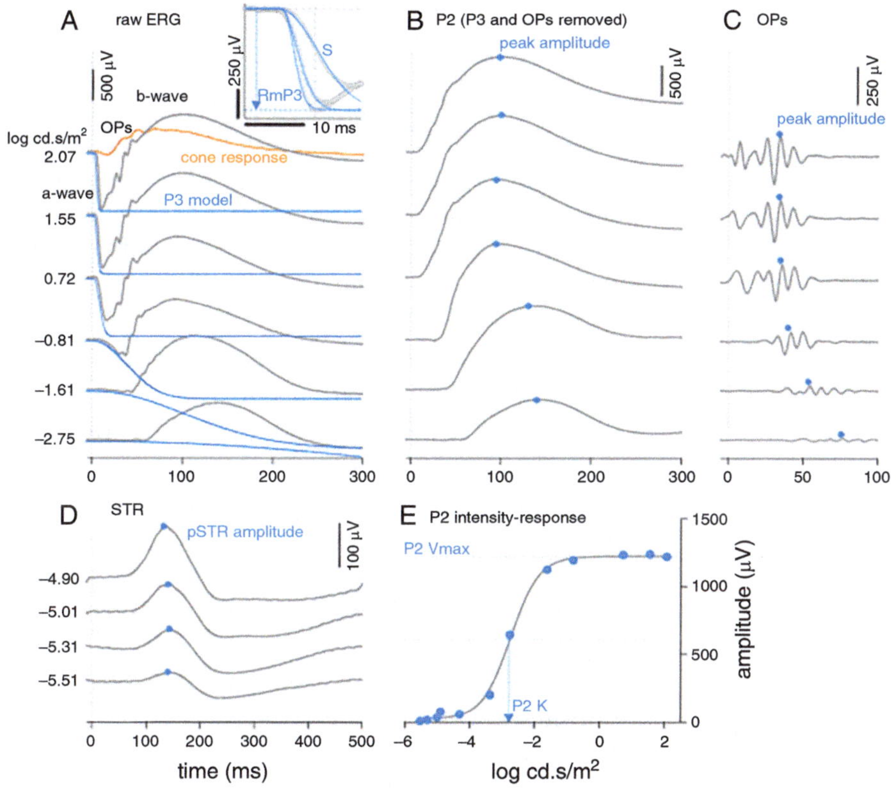


**Fig. 2** Analysis of the rodent electroretinogram (ERG). (**a**) Raw ERG waveforms (from a mouse) for moderate to bright flashes. Rats produce similar sequence of responses. The photoreceptor response (a-wave) can be modelled using a delayed Gaussian function (P3, blue traces), which is fit to the brightest light levels (inset). This model returns a maximum amplitude (RmP3) and a slope parameter, which indicates sensitivity to light (S). The cone response (orange trace) can be isolated by using a very short interstimulus interval (0.5 s) between subsequent bright flashes. (**b**) The bipolar cell response (P2) can be isolated by subtracting the P3 model from the raw ERG and low-pass filtering (<50 Hz) to remove oscillatory potentials (OPs) and high-frequency noise. Peak P2 amplitudes and implicit times can be obtained at each light level. (**c**) Oscillatory potentials can be isolated by band-pass filtering the ERG waveform (50–180 Hz) and taking peak amplitudes and implicit times. (**d**) The dimmest light levels elicit the scotopic threshold response (STR); the positive component provides a measure of ganglion cell function. (**e**) Peak amplitudes can be plotted as a function of light level and modelled using a saturating function to return a maximum amplitude (P2 Vmax) and a sensitivity to light (1/K)

19. The dark-adapted ERG elicited with bright light levels (Table 1) contains contributions from both the rod and cone pathways. The long refractory time of the rods can be leveraged using repeated stimuli with short interstimulus intervals to elicit subsequent responses that are cone dominated. Cone responses can be quantified by their peak amplitudes and implicit times (Fig. 2a orange trace).

### 3.2 Optical Coherence Tomography for Retinal Morphology Assessment

1. Check to ensure that the animal is deeply anesthetized and there is no excessive movement. Place the animal on the rodent platform, and add a drop of ocular lubricant (i.e., Systane) to smooth the corneal surface. Orient the animal so that the optical axis is roughly aligned with the axis of the camera.
2. Twist the tip of a Kimwipe and gently sweep first along the upper eyelids; this moves long eyelashes away from the cornea. Then sweep the tip gently along the lower lid margin to remove any excess ocular lubricant. Repeat this step if image quality needs to be improved.
3. While the first eye is being imaged, keep the other eye moist with a large drop of hydrating gel.
4. Attach the 25D lens to the front of the objective for mice. For rats, this lens is not necessary.
5. Enter animal identification details. Choose the OCT imaging option and the scan pattern (*see* **Notes 3** and **4**, e.g., 7.9 × 6.6 mm, 768 A-scans per B-scans, 121 B-scans evenly spaced in the vertical dimension with averaging of 5 images each B-scan as per Fig. 3a).
6. Align the optical axis of the eye along the objective lens. Move the objective lens to within 10–15 mm from the eye. Adjust focus until the large blood vessels and the nerve fibers are in focus (Fig. 3a). Adjust objective position (up/down, side to side, as well as pitch) to (i) center the optic nerve in the image, (ii) so that the live OCT scan is horizontally flat and (iii) the image quality is optimized, and then acquire the volume.
7. During acquisition, check that image quality is consistently high. Quality of 30 dB or greater returns images that are of quality that allows quantification.
8. Image analysis can be undertaken via automated segmentation or manual segmentation. Manual segmentation by a masked observer can be undertaken using Fiji software (https://fiji.sc/). B-scans through the center of the optic nerve are analyzed by manually tracing the structural boundaries of interest, for example, retinal nerve fiber layer (inner limiting membrane) and Bruch's membrane (Fig. 3b). From these boundaries, total retinal thickness can be quantified. As the retina thins toward the optic nerve, it is best to avoid these areas.
9. Alternatively, use the Posterior Pole tab to access the built-in (HEYEX) automated segmentation. Choose the layer of interest and export the thickness values (Fig. 3c).

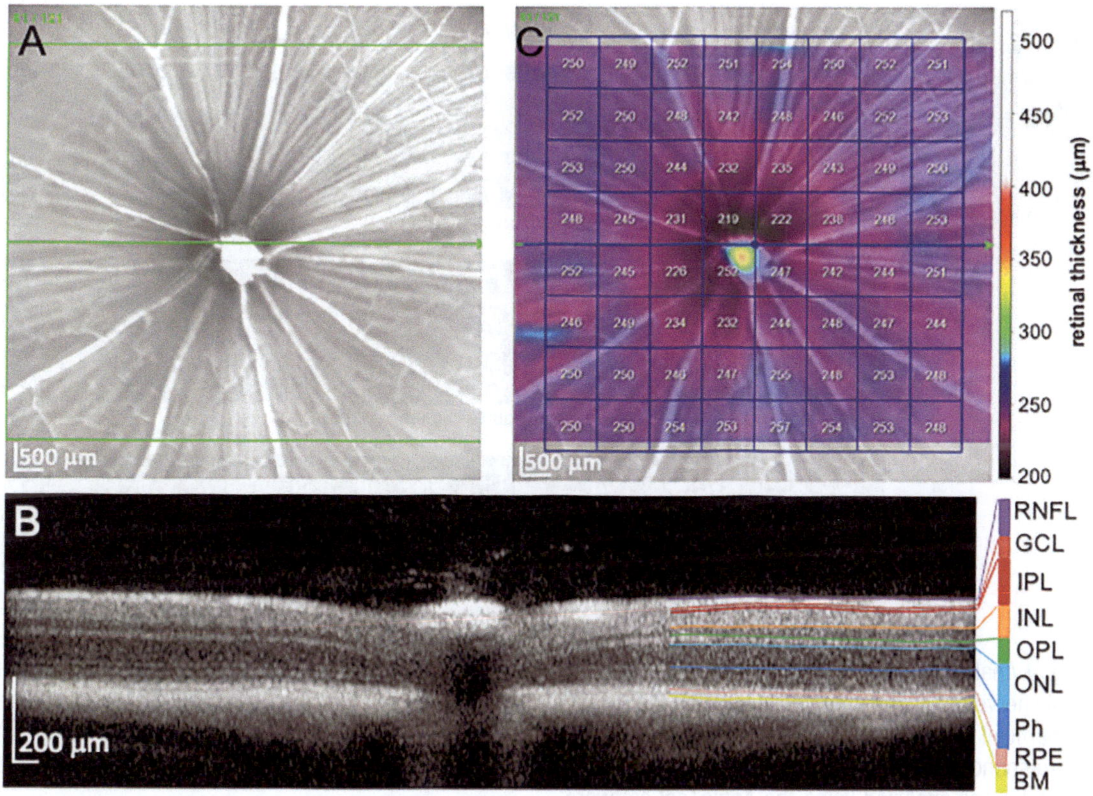


**Fig. 3** Analysis of rodent structural OCT. (**a**) A volume scan (7.9 × 6.6 mm, 121 B-scans) centered on the optic nerve head. The green box indicates the extent of the scan area. The center line indicates the middle B-scan, as shown in panel b. (**b**) Manual segmentation can be done by a masked observer to demarcate boundaries of the retinal nerve fiber layer (RNFL), ganglion cell layer (GCL), inner plexiform layer (IPL), inner nuclear layer (INL), outer plexiform layer (OPL), outer nuclear layer (ONL), photoreceptors (Ph), retinal pigment epithelium (RPE), and Bruch's membrane (BM). (**c**) HEYEX provides in-built segmentation and data analysis to return thickness measurements for specific retinal layers

### *3.3 Optical Coherence Tomography Angiography for Retinal Vascular Assessment*

1. Choose the OCTA imaging option and the desired scan pattern (Fig. 4a shows a 20° ×10° scan, with 512 A-scans for each of 256 B-scan, with averaging of 4 for each B-scan).
2. Select a consistent region of the retina to scan (e.g., away from the optic nerve, Fig. 4a; also *see* **Note 5**).
3. Adjust objective position (up/down, side to side, as well as pitch) to (i) consistently place the optic nerve at the edge of the image, (ii) so that the live OCT scan is horizontally flat and (iii) the image quality is optimized, and then acquire the volume.
4. During acquisition, check that image quality is consistently high. Quality of 30 dB or greater returns images that can be analyzed reliably.
5. Prior to saving the volume, inspect the preview to ensure there are no movement artifacts or regions where vessels were not detected.

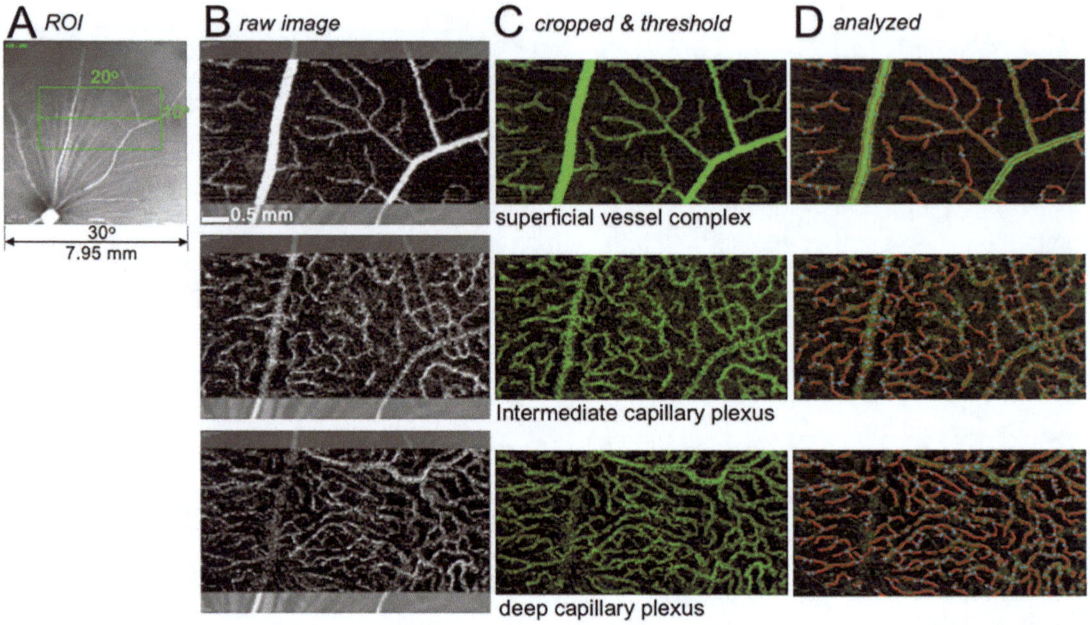


**Fig. 4** Analysis of rodent OCT angiography. (**a**) A volume scan (7.9 × 6.6 mm, 121 B-scans) centered on the optic nerve head. The green box indicates the extent of the scan area. (**b**) Examples of saved .tiff images of the superficial vessel complex (top), intermediate capillary plexus (middle), and deep capillary plexus (bottom). (**c**) Corresponding images cropped and thresholded using the AngioTool software. (**d**) Selected overlays of retinal vessel areas analyzed using the AngioTool software

6. To analyze the OCTA image, open the volume in the built-in analysis algorithm (HEYEX), and check that the segmentation is correct for the retinal layers.
7. Use the OCT angiography tab to save the superficial vessel complex (SVC), intermediate capillary plexus (ICP), and deep capillary plexus (DCP) images as separate Tag Image File Format (.tiff) image files (Fig. 4b). Ensure that all overlays are removed prior to exporting the images.
8. Use Fiji software to crop edges of the .tiff image files leaving only the region containing the OCTA signal (Fig. 4c). Consistent cropping can be achieved by creating a stack of adjacent images for each eye.
9. Use the AngioTool software to return vessel area [15]. Open one image at a time in the AngioTool software. Set diameter (e.g., 40), set remove small particles (e.g., 500), and set fill holes (e.g., 500). Enter calibration details "Distance in Pixels" (e.g., 56 pixel) and "Distance in mm" (e.g., 0.2 mm) to return vessel parameters in mm.
10. Run analysis which will save the thresholded image (Fig. 4c) with the selected overlays (Fig. 4d) along with a results Excel file (.xlsx) in the parent folder.

## 4 Notes

1. General anesthesia in rodents can induce a reversible fluid opacification of the murine lens. This can be avoided by keeping the cornea well hydrated using a viscous gel (e.g., GenTeal).
2. The scotopic threshold response is a sensitive indicator of early retinal injury in rodent models of diabetes [16]. This component is highly sensitive to background light and requires absolute dark adaptation [17] and stable anesthesia. Increased signal averaging of 100–200 signals may be required [14].
3. OCT/OCTA scans can be set to either the vertical or horizontal direction. Image capture is faster when the scan direction is set perpendicular to the larger blood vessels.
4. With longitudinal assessment of retinal structure, the follow-up mode can be used by setting the initial scan as the reference. Examining the same retinal area substantially reduces variability.
5. OCTA scans are best away from the optic nerve as image segmentation of retina structural boundaries closer to the optic nerve is difficult, thus making the determination of vessel plexi bases more prone to error in these areas.

### References

1. Grillo SL, Koulen P (2015) Psychophysical testing in rodent models of glaucomatous optic neuropathy. Exp Eye Res 141:154–163. https://doi.org/10.1016/j.exer.2015.06.025
2. Leinonen H, Tanila H (2018) Vision in laboratory rodents-tools to measure it and implications for behavioral research. Behav Brain Res 352:172–182. https://doi.org/10.1016/j.bbr.2017.07.040
3. Pinto LH, Enroth-Cugell C (2000) Tests of the mouse visual system. Mamm Genome 11(7):531–536. https://doi.org/10.1007/s003350010102
4. Frishman L (2006) Origins of the electroretinogram. In: Heckenlively JR, Arden GB (eds) Principles and practice of clinical electrophysiology of vision. The MIT Press, Cambridge, pp 139–184
5. Granit R (1933) Components of the retinal action potential in mammals and their relations to the discharge in the optic nerve. J Physiol 77:207–238
6. Weymouth AE, Vingrys AJ (2008) Rodent electroretinography: methods for extraction and interpretation of rod and cone responses. Prog Retin Eye Res 27(1):1–44. https://doi.org/10.1016/j.preteyeres.2007.09.003
7. Qiu F, Zhu M, Le YZ (2019) Noninvasive diagnosis of regional alteration of retinal morphology and structure with optical coherence tomography in rodents. Adv Exp Med Biol 1185:151–155. https://doi.org/10.1007/978-3-030-27378-1_25
8. Mills SA, Jobling AI, Dixon MA, Bui BV, Vessey KA, Phipps JA, Greferath U, Venables G, Wong VHY, Wong CHY, He Z, Hui F, Young JC, Tonc J, Ivanova E, Sagdullaev BT, Fletcher EL (2021) Fractalkine-induced microglial vasoregulation occurs within the retina and is altered early in diabetic retinopathy. Proc Natl Acad Sci U S A 118(51). https://doi.org/10.1073/pnas.2112561118
9. Zhao D, Nguyen CT, Wong VH, Lim JK, He Z, Jobling AI, Fletcher EL, Chinnery HR, Vingrys AJ, Bui BV (2017) Characterization of the circumlimbal suture model of chronic IOP elevation in mice and assessment of changes in gene expression of stretch sensitive channels. Front Neurosci 11:41. https://doi.org/10.3389/fnins.2017.00041

10. Zhao D, Wong VHY, Nguyen CTO, Jobling AI, Fletcher EL, Vingrys AJ, Bui BV (2019) Reversibility of retinal ganglion cell dysfunction from chronic IOP elevation. Invest Ophthalmol Vis Sci 60(12):3878–3886. https://doi.org/10.1167/iovs.19-27113
11. Lamb TD, Pugh ENJ (1992) A quantitative account of the activation steps involved in phototransduction in amphibian photoreceptors. J Physiol 449:719–758
12. Naka KI, Rushton WA (1966) S-potentials from colour units in the retina of fish (cyprinidae). J Physiol 185(3):536–555
13. Wachtmeister L (1998) Oscillatory potentials in the retina: what do they reveal. Prog Retin Eye Res 17(4):485–521
14. Bui BV, Fortune B (2004) Ganglion cell contributions to the rat full-field electroretinogram. J Physiol 555(Pt 1):153–173
15. Zhao D, He Z, Wang L, Fortune B, Lim JKH, Wong VHY, Nguyen CTO, Bui BV (2020) Response of the trilaminar retinal vessel network to intraocular pressure elevation in rat eyes. Invest Ophthalmol Vis Sci 61(2):2. https://doi.org/10.1167/iovs.61.2.2
16. Kohzaki K, Vingrys AJ, Armitage JA, Bui BV (2013) Electroretinography in streptozotocin diabetic rats following acute intraocular pressure elevation. Graefes Arch Clin Exp Ophthalmol 251(2):529–535. https://doi.org/10.1007/s00417-012-2212-4
17. Saszik SM, Robson JG, Frishman LJ (2002) The scotopic threshold response of the dark-adapted electroretinogram of the mouse. J Physiol 543(Pt 3):899–916

# Chapter 5

# A Guide to Genome-Wide Association Study Design for Diabetic Retinopathy

Rajya L. Gurung, Kathryn P. Burdon, and Bennet J. McComish

## Abstract

Diabetic retinopathy (DR) is the most common microvascular complication related to diabetes. There is evidence that genetics play an important role in DR pathogenesis, but the complexity of the disease makes genetic studies a challenge. This chapter is a practical overview of the basic steps for genome-wide association studies with respect to DR and its associated traits. Also described are approaches that can be adopted in future DR studies. This is intended to serve as a guide for beginners and to provide a framework for further in-depth analysis.

**Key words** Diabetic retinopathy, Genome-wide association study, Genetics, Methods, Protocols

## 1 Introduction

Diabetic retinopathy (DR), the most common microvascular complication of diabetes, is one of the leading causes of blindness in the working-age population [1]. Several clinical risk factors have been linked to DR, including duration of diabetes, glycemic control (HbA1c), hypertension (HTN), nephropathy, and hyperlipidemia [2]. Like many other complex diseases, DR is caused by a combination of genetic and environmental risk factors [3]. The advent and increasing affordability of high-throughput genotyping in the last decade have enabled comprehensive genomic investigations of several blinding eye conditions using genome-wide association study (GWAS) designs [4, 5]; however, DR-related GWAS have yet to realize their full potential [6, 7]. One limitation is that the various classifications and severity scales used lead to heterogeneity of the DR phenotype and make combining or comparing studies difficult. This aspect requires careful consideration when designing and analyzing genome-wide scans for this spectrum of disease. This chapter provides a practical overview of protocols, methods, tools, and pipelines for the conduct of GWAS of DR and related traits.

Guei-Sheung Liu and Jiang-Hui Wang (eds.), *Diabetic Retinopathy: Methods and Protocols*, Methods in Molecular Biology, vol. 2678, https://doi.org/10.1007/978-1-0716-3255-0_5,

First, it describes various traits related to DR that can be carried forward to GWAS. Further, it outlines the steps underlying sample processing, quality control, imputation, adjustment for population stratification, and association analyses. Where applicable, we point to key references, software programs, and texts for the reader. Methods and tools available for GWAS are an evolving and dynamic field of active research with diverse opinions on approaches. Advanced researchers in the field of GWAS and DR will be familiar with the approaches and may suggest alternatives. The protocols presented herein are targeted at researchers seeking the basics of GWAS in the context of DR or other diseases. For ease of understanding, most of the commands to execute the computational steps are provided. Except for the section on data cleaning using GenomeStudio (Version 2.0, Illumina), all commands are performed in a UNIX-like environment. The R codes (*provided at the end of the chapter after the "Notes" section*) can be used in R/RStudio (http://www.rstudio.com/).

## 2 Materials

### 2.1 Participant Information (See Note 1)

1. Clinical examination conducted by ophthalmologists, optometrists, and trained eye healthcare workers.
2. Medical records (digital or handwritten).
3. Fundus photographs, optical coherence tomography (OCT) images.
4. Participant recruitment questionnaire forms and informed consent documentation.

### 2.2 Samples for Genomic DNA

1. Peripheral venous blood (collected in EDTA tube).
2. Saliva samples (e.g., Oragene Saliva kit, DNA Genotek).

### 2.3 DNA Extraction

1. Kit or method of your choice for DNA extraction, appropriate to the sample type collected.
2. Kit or method for fluorometric assessment of DNA concentration.
3. Spectrophotometer for assessment of DNA quality.

### 2.4 Genotyping, Quality Control, and Imputation

1. SNP array of choice. Methods for genotyping are not covered here. Third-party providers can be used.
2. GenomeStudio (Version 2.0, Illumina).
3. PLINK 1.9/2.0 [8, 9].
4. Michigan Imputation Server [10].

### 2.5 Genome-Wide Association Analysis and Data Visualization

1. PLINK 1.9/2.0 [8, 9].
2. qqman package [10].
3. GWASTools, an R/Bioconductor package [11, 12].
4. LocusZoom [13].
5. FUMA v1.3.6 [14].
6. Linux-based operating system/high-performance computing cluster with large data storage capacity is preferred if there are thousands of samples, but for small- to moderate-sized studies, one can perform these steps on a personal desktop with at least 16 GB of RAM.
7. R (https://www.r-project.org/) for data visualization.
8. Python.
9. BCFtools [15] (https://samtools.github.io/bcftools/bcftools.html).

## 3 Methods

### 3.1 Phenotype Categorization

Table 1 presents phenotypes and study design approaches relevant for the genetic analysis of DR. Obtaining high-quality phenotypes is the first important but often under-appreciated step before carrying out a GWAS [16]. Defining the disease or phenotype of interest as accurately and specifically as possible is necessary to minimize genetic and environmental heterogeneity. Indeed, phenotype resolution is likely a major determinant of the success or failure of GWAS to date [17]. For traits with well-defined phenotypes with large effect sizes, small cohorts have sufficed for performing a "successful" GWAS [18]. In contrast, for a complex disease such as DR, the phenotype is less precise in the sense that there are several subtypes of DR with overlapping features [19]. Briefly, the early stages of DR are labeled as non-proliferative DR (NPDR), which can progress to vision-threatening stages of proliferative DR. Diabetic patients may also develop diabetic macular edema (DME), affecting the central part of the retina, which causes much of the vision loss due to DR [19]. DR is subject to several confounding factors such as duration of diabetes, diabetes control (HbA1c), nephropathy status, HTN, age, and sex [20]. Thus, this phenotypic heterogeneity of DR and DME makes standardized phenotype classification difficult. Consequently, careful thought should be given to the type of DR phenotype to be collected before a GWAS experiment can be designed. Although the typical samples sizes that are required to discover novel genome-wide significant risk factors in complex diseases are in the thousands, there are some precedents of strong genetic

**Table 1**
**Traits related to diabetic retinopathy that can be used for genome-wide association study**

| Trait | Type | Case description | Control description |
|---|---|---|---|
| DR | Binary | Diabetic patients with any grade of diabetic retinopathy | Diabetic patients without DR |
| Proliferative DR | Binary | Diabetic patients with any feature of proliferative retinopathy | Diabetic patients without DR |
| DME | Binary | The presence of any clinically significant macular edema irrespective of vision and diabetic retinopathy status | Diabetic patients without DR |
| DME with significant vision loss | Binary | The presence of clinically significant macular edema with vision <70 ETDRS letters | Diabetic patients without DR |
| Sight-threatening DR | Binary | The presence of severe non-proliferative DR or proliferative DR or center involving DME | Diabetic patients without DR |
| Change in BCVA | Continuous | BCVA converted to logMAR scale or ETDRS scale (*see* **Note 2**) | – |
| CMT in DME | Continuous | CMT measured by optical coherence tomography (*see* **Note 3**) | – |
| HbA1c | Continuous | HbA1c measured by standard technique at the time of recruitment | – |
| *Extreme phenotype approach* | | | |
| DR extreme | Binary | Diabetic patients with severe or rapid onset DR | Diabetic patients who do not develop DR ≥15 years after diagnosis of DM [3] |

Abbreviations: *BCVA* best-corrected visual acuity, *CMT* central macular thickness, *DM* diabetes mellitus, *DME* diabetic macular edema, *DR* diabetic retinopathy, *ETDRS* early treatment diabetic retinopathy study, *VEGF* vascular endothelial growth factor

effects detected with small but phenotypically homogenous samples [21, 22]; however, to date, this has not proven to be the case for DR phenotypes.

The other approach that can be adopted for GWAS of DR is to utilize an intermediate or proxy phenotype that is correlated to the target phenotype of interest—also known as an endophenotype. Endophenotypes aid in resolving genetic heterogeneity commonly found in complex diseases [23, 24]. The different stages of DR are based mostly on Early Treatment Diabetic Retinopathy Study (ETDRS), a well-established set of clinical criteria for the classification of DR [19]. However, there may be variability between clinicians and different study sites in how these criteria are used. An endophenotype can be used to increase the homogeneity of the cohort compared to an endpoint phenotype. Endophenotypes are quantitative traits that are heritable, measurable in the whole population regardless of disease status, and genetically correlated with the disease of interest. True endophenotypes for DR are limited; however, central macular thickness (CMT) is a reasonable example, particularly in relation to diabetic macular edema. Glycosylated hemoglobin (HbA1c) may also be relevant to all diabetic complications. Other quantitative traits may also be useful depending on the context. For example, change in best-corrected visual acuity (BCVA) over time is likely to be useful as a proxy for DR severity, treatment response, or disease progression when assessed in diabetic populations. Another useful approach for GWAS of DR is to use extreme phenotypes [25]. There are several clinical risk factors for DR; however, some patients develop DR despite good glycemic control, whereas others with poor glycemic control are spared [26]. The Joslin 50-year medalist study is one study that demonstrates a significantly large population who remain free of diabetic complications despite a considerable duration of DM [27]. Thus, extreme phenotype analysis compares diabetic patients without any feature of DR despite a long duration of DM and poor glycemic control against patients with DR (ideally patients with DR despite a short duration of DM and good diabetes control) [3, 27]. This approach helps in reducing not only the cost of genotyping/sequencing by narrowing the number of patients but also phenotype heterogeneity, without losing power [28, 29]. This approach is particularly beneficial when risk genotypes are rare in the population and effect size is relatively small [30]. As such, numerous traits are available for the study of DR in the context of a GWAS.

### 3.2 Study Design

Another important aspect for a successful GWAS is the study design. Similar to most GWAS designs for complex diseases [31], for DR, the commonly used approach is a case-control design [32–35]. Overall, the case-control study design compares the frequency of alleles and polymorphic sites between the cases and the controls. However, as there are several subtypes of DR with considerable overlap between them, it is critical to carefully match cases and controls to avoid confounding and spurious associations. Consideration should be given to using control participants with HbA1C > 7.5% and with a duration of diabetes of at least 10 years, but without DR [27] to minimize this risk of misclassifying pseudo-controls who might develop the disease later; however, stringent control definitions such as this will limit the ability to recruit sufficient numbers of diabetic controls. Another common approach is a quantitative study design utilizing endophenotypes. Several studies have demonstrated that a quantitative study design is statistically more powerful than a case-control design for detecting genetic effects [16, 36]. Quantitative traits also allow researchers to provide an outcome that is clinically easier to interpret [37]. Finally, it is imperative that for GWAS of DR, appropriate adjustments for confounding factors be made. As highlighted above, several landmark epidemiological studies have identified duration of diabetes, hyperglycemia, HTN, nephropathy, and hyperlipidemia as the major risk factors associated with DR [38, 39]. These factors need to be properly adjusted for in the analysis to minimize their effects.

### 3.3 Power of the Study

Estimating a sufficient sample size to achieve adequate statistical power is critical in the design stage of genetic association studies [40]. The many GWAS on DR published to date have yielded very few significant associations and none with clear replication between studies [6, 41]. One possible explanation for the varying outcomes is that the risk alleles have small to modest effect sizes, implying a need for larger phenotypically homogenous sample cohorts or large meta-analyses between cohorts to increase power. There are several tools available for power and sample size calculations for genetic association studies. We commonly use Genetic Power Calculator developed by Purcell et al. [42] (https://zzz.bwh.harvard.edu/gpc/) for our sample size/power calculation. This site provides automated power analysis for different test models. The steps for case-control (additive model) and quantitative design are given below.

### 3.4 Case-Control Design

1. Go to https://zzz.bwh.harvard.edu/gpc/ (this will lead to the home page).
2. Under modules, click "Case-control for discrete traits" and specify the parameters as below.
3. High-risk allele frequency (A) = frequency of allele being tested (*see* **Note 4**).
4. Prevalence = prevalence of disease.
5. Genotype relative risk Aa = odds ratio of a single copy of the risk allele "A" (*see* **Note 5**).
6. Genotype relative risk AA = odds ratio for two copies of the risk allele (*see* **Note 6**).
7. D-prime = this is a measure of linkage disequilibrium (LD) between the true causative variant and the marker being genotyped, usually set to 1.0, indicating strong LD when using dense genotype arrays with imputation.
8. Marker allele frequency (B) = same as the allele frequency "A" when LD is strong (*see* **Note 7**).
9. Number of cases = how many cases are available.
10. Control-case ratio = how many controls compared to cases (i.e., 1, the same number of cases and controls; 4, four times more controls than cases).
11. Unselected controls = leave unchecked (default), if controls have been screened for DR and selected to have diabetes (*see* **Note 8**).

12. User-defined type 1 error rate = the *p*-value threshold to indicate significance (commonly use $5 \times 10^{-8}$ for GWAS).
13. User-defined power = power you want to detect an association, typically 80% (0.8).
14. Click "Process."
15. Output: The different tables give the power of the study at a range of *p*-values (alpha) for different genetic models.

### 3.5 Quantitative Trait

1. Under modules, click "QTL Association for sibships" and specify the parameters as below.
2. Total QTL variance = percentage of phenotypic variance explained by the locus given as $R^2$ value in plink.
3. Check "No dominance" (this will calculate only additive effect).
4. QTL increaser allele frequency = allele frequency for the biallelic QTL.
5. Marker M1 allele frequency = same as QTL increaser allele above (assuming LD = 1).
6. Linkage disequilibrium = 1.
7. Sibling correlation = 0.5 if participants are unrelated.
8. Sample size = number of samples available.
9. Sibship size = Singleton.
10. User-defined type 1 error rate = the p-value threshold (commonly use $5 \times 10^{-8}$ for GWAS).
11. User-defined power = power you want to detect an association, e.g., 80% (0.8).
12. Process.

### 3.6 Collection of Samples for Genomic DNA

1. A venous blood sample obtained from each participant by a trained venipuncturist. Two 9 mL Vacuette EDTA tubes of venous blood are recommended. They should be stored at 4 °C prior to use or −80 °C for long-term storage (*see* **Note 9**).
2. Alternatively, DNA can be extracted from other tissues, including saliva. Several collection kits for saliva suitable for DNA extraction are available.

### 3.7 DNA Extraction from Blood or Saliva

DNA quality and quantity are prerequisites for genotyping success [43]. Differences in DNA quality can result in differences in the frequency of missing genotype calls, thus inflating type 1 error [44]. Conversely, poor-quality DNA leading to "noise" in the data may increase false-negative results. Therefore, rigorous standards for the extraction and purification of genomic DNA should be established to minimize any genotyping errors. While many kits/protocols are available for DNA extraction [43] and quantitation [45], we target a minimum concentration of 50 ng/μL. A 260: 280 ratio of ≈1.8 and 260:230 ≈ 1.8–2.2 are generally accepted as "pure" for DNA (https://dna.uga.edu/wp-content/uploads/sites/51/2019/02/Note-on-the-260_280-and-260_230-Ratios.pdf).

### 3.8 Genotyping

Several genotyping platforms are available (reviewed in [46]). GWAS using either of these platforms has been equally successful in the search for genetic risk factors for common, complex diseases. The methods described in this chapter utilize the Illumina SNP genotyping arrays with which we have experience but are applicable to any high-density dataset.

1. Ship DNA samples to genotyping centers according to each facility's instructions and best practices. Ensure that the samples are labeled properly and correctly. If a case-control design, ensure cases and controls are randomly distributed across plates and arrays to avoid batch effects that create spurious associations.
2. Post-genotyping, the data provided are presented in Table 2.

### 3.9 Genotype Calling and Data Cleaning

The following section describes in detail genotype calling and data cleaning steps using Illumina-specific software GenomeStudio (Version 2.0), designed for the Windows operating system. The Illumina genotyping arrays (http://www.illumina.com/techniques/microarrays) use probes designed to hybridize to specific regions of genomic DNA, followed by extension using chemically labeled nucleotides. The probe extension binds either red or green fluorescent agents. An Illumina array scanner detects these red and green fluorescent intensities for each SNP locus to produce genotype data. GenomeStudio converts red and green color channel

**Table 2**
**Files provided from the genotyping center (*see* Note 10)**

| File type | File description | Number of files |
|---|---|---|
| Intensity data file folders (*.idat) | Folder labeled with array serial numbers and containing red and green *.idat files for each sample | There will be many *.idat folders with unique serial numbers in a parent folder named "idats" |
| Manifest file (*.bpm) | Files containing bead IDs and SNP information for each marker on array | Single file |
| Cluster file (*.egt) | Contains the genotyping cluster positions | Single file (OPTIONAL) |
| Sample sheet (*.csv) | Contains user-specified information on each sample genotyped | Single file (*see* **Note 11**) |

signals into genotype calls by automated clustering. However, the default clustering algorithm can fail to identify valid clusters and can also assign the wrong genotype to samples due to abnormal intensity patterns. This can be addressed by manually reviewing and recalling SNPs to increase the reliability, confidence, and overall quality of the data. This genotype calling and data cleaning process in GenomeStudio (Genotyping module) is given below (*see* **Note 12**). The steps outlined below have been adopted from Guo et al. [47] and Infinium Genotyping Data Analysis [**Tech Note**, (https://www.illumina.com/Documents/products/technotes/technote_infinium_genotyping_data_analysis.pdf)] along with recommendations gathered from our own experience. Readers are recommended to refer to the figures provided in the **Tech Note** for better understanding of the steps detailed below. Various metrics are commonly used, and there is some level of subjectivity, particularly when manually editing the clusters.

1. Download GenomeStudio from https://sapac.support.illumina.com/array/array_software/genomestudio/downloads.html.
2. Create a common project folder with all the files in Table 2 (*see* **Note 13**).
3. Load the data into GenomeStudio using Sample Sheet.
   - (a) Select File → New Project → Genotyping.
   - (b) This will open up a GenomeStudio Project Wizard → Click "Next."
   - (c) Under "Projects Repository," select the project folder created in **step 2**.
   - (d) Under "Project Name," give the project a meaningful name in the "Create" section.
   - (e) Click "Next."
   - (f) Select "Use sample sheet to load sample intensities."
   - (g) Under "Sample Sheet," click Browse, and select your sample sheet file (*.csv) in the parent folder created in **step 2**.
   - (h) Under "Data Repository," click Browse, and specify your parent intensity data folder (e.g., idats) within the project folder created in **step 2**.
   - (i) Under "Manifest Repository," click Browse, and specify your manifest file (*.bpm) in the project folder created in **step 2**.
   - (j) Click "Next."
   - (k) Check the box "Import cluster positions from a cluster file" (*see* **Note 14**).

(l) Under "Cluster File," click Browse, and specify your cluster file (*.egt) in the project folder created in **step 2**.

(m) Under "Project Creation Actions," check "Calculate Sample and SNP Statistics."

(n) Ensure "Gen Call Threshold" is set at 0.15 (*Default*), and click "Finish" to create the project (*see* **Note 15**).

(o) Click Save once the project is created (*see* **Note 16**).

(p) Also, do Files → Save Project Copy As → Give a suitable file name, e.g., RawFile (*see* **Note 17**).

4. Loading the data into GenomeStudio without using Sample Sheet (*Alternative* to **step 3**).

   (a) Follow **steps (a)**–**(e)** from above.

   (b) Select "Load sample intensities by selecting directories with intensity files" → Select the project folder created in **step 2**. Click "Next."

   (c) Under "SNP Manifest," click Browse, and specify your manifest file "(*.bpm)" in the project folder created in **step 2**.

   (d) Under "Data Repository," click "Browse," and specify your parent intensity data folder (e.g., idats) within the project folder created in **step 2**.

   (e) Under "Directories in Repository," select one or multiple idat folders with serial numbers, and add them to the "Selected Directories" on the right side.

   (f) Follow **steps (k)** to **(p)** from Subheading 3.

5. Checking overall experimental quality (*see* **Note 18**).

   (a) Analysis → View Controls Dashboard.

   (b) A new window will open with all the control graphs. Details of how to evaluate the control graphs are provided in https://support.illumina.com/content/dam/illumina-support/courses/eval-inf-controls/story_content/external_files/Infinium_Controls_Training_Guide.pdf.

While this is not the entirety of the quality control (QC) process, it is an important step. It helps to identify sample outliers and the root causes of assay failure. It is always recommended to check the control dashboard even if the call rate looks good, but one should never draw conclusions by looking at a single control. It is recommended to evaluate controls in combination with each other and with other quality metrics in GenomeStudio. For example, if samples failing in the "Control graphs" also have poor call rates, then we are more confident to exclude those samples or re-genotype them. Further, one should never conclude from absolute intensity levels of individual samples; it is recommended to look at the relative intensity level (range of intensity). There are no specific thresholds for high, medium, or background intensities; it is variable between projects, bead chips, experiments, and even scanners.

6. Sample QC (all steps in Samples Table console).
   (a) From the Samples Table, arrange by "Call Rate" either descending or ascending, highlight all samples with a call rate of <0.95, right-click, and "Exclude Selected Samples" (*see* **Note 19**).
   (b) After the samples are excluded, you will be prompted to update sample and SNP statistics; click "No" (*see* **Note 20**).
   (c) To improve the overall clustering, it is recommended to re-cluster after excluding samples with poor call rate: Highlight all the SNPs in the SNP table → right mouse click → Cluster All SNPs. The resulting clusters should be cleaner.
   (d) Often sex is not given in the sample sheet, or incorrect sex will be assigned. To cross-check this discrepancy, in the Samples Table console, click the "Column chooser" icon from the toolbar → Hidden Columns → drag "Gender Est" to Displayed Columns → OK. Gender Est column will now be displayed in the Samples Table console, but all samples will have "Unknown" against their gender estimates.
   (e) Select all the samples. The selected samples will be highlighted blue. Then click the right mouse button → Estimate Gender for Selected Sample → Would you like to populate the Gender column as well? → Select "No."
   (f) The sex of the samples should match between the "Gender Est" column and the "Gender" column. Any mismatch means data entry error or sample mix-up (*see* **Note 21**).
7. QC of chromosome 0 SNPs (SNP Table console).

   Chromosome 0 SNPs in the GenomeStudio project are problematic probes (these have no valid mapping, map to multiple locations, are no longer used in the newer version of the array, or map to contigs that are not yet located on a particular chromosome in the latest genome assembly) and thus should be omitted (zeroed) before analysis.

   (a) From the "SNP Table" toolbar, select "Clear filter" to ensure no other filters are active.
   (b) From the "SNP Table," select "Filter rows" → in Column, select "Chr" → in Operation, select "=" → in Value, write "0," and then click "==>" and then "OK," i.e., Filter = [("Chr" = 0)].

(c) The above step will output all the SNPs in chromosome 0.

(d) Zero all the SNPs, which is done by right-clicking on a SNP (a row) in the "SNP Table" and by selecting "Zero Selected SNP" OR by pressing the F2 keyboard button.

8. QC of haploid SNPs {X, Y, and mitochondrial (MT) SNPs}.

   (a) From the "SNP Table" toolbar, select "Clear filter."

   (b) From the "SNP Table," select "Filter rows" → in Column, select "Chr" → in Operation, select "=" → in Value, write "X," and then click "==>"; next, select "!=" in Operation → in Value, write "XY," and then click "==>" and then "OK."

   (c) The above step will output all the SNPs on chromosome X excluding pseudo-autosomal SNPs (XY), i.e., Filter = [("Chr" = X) AND ("Chr"! = XY)].

   (d) In the "Samples Table," sort samples by gender and select all the male samples. These should now be highlighted in yellow in the SNP graph. All female samples should be colored according to their genotype.

   (e) In the "SNP Table," select the "AB Freq" column by clicking on the column header. Sort this column in descending order by clicking on the "Sort column (Descending)" button located at the top of the table.

   (f) Manually check the SNPs in descending order, and look for AB clusters with male subjects represented by yellow dots. Because males should not appear in the AB cluster on the X chromosome, this manual check identifies mis-clustered X chromosome SNPs. Zero these SNPs with any yellow dot in the AB cluster (*see* **Note 22**).

   (g) From the "SNP Table" toolbar, select "Clear filter."

   (h) As detailed above, in the "SNP Table," change Filter = [("Chr" = Y) AND ("Chr"! = XY)].

   (i) The above step will output all the SNPs on chromosome Y excluding pseudo-autosomal SNPs (XY).

   (j) In the "Samples Table," select all the female samples. These should now be highlighted in yellow in the SNP graph. All male samples should be colored according to their genotype.

   (k) Sort the SNPs in the "SNP Table" by "Call Freq" in descending order on the Y chromosome. The call rate for females on the Y chromosome should be at or near 0%, i.e., there should be no yellow dots (female samples) in the SNP graph for Y chromosome. Zero the SNPs with females called on the Y chromosome.

(l) From the "SNP Table" toolbar, select "Clear filter."

(m) In the "SNP Table," change Filter = [("Chr" = MT)].

(n) Sort the SNPs in "SNP Table" by "AB freq" in descending order for mitochondrial DNA SNPs. These SNPs are haploid and therefore should show only AA and BB clusters, although some AB clusters may be present due to heteroplasmy. Zero mitochondrial DNA SNPs with high AB cluster frequency.

9. QC of autosomal SNPs.

There are various parameters under which autosomal SNPs are assessed. The parameters vary according to the project and the type of genetic analysis to be performed. The main QC parameters for which autosomal SNPs are filtered on are described below.

(a) From the "SNP Table" toolbar, select "Clear filter."

(b) In the "SNP Table," change Filter = [("Chr"! = X) AND ("Chr"! = Y) AND ("Chr"! = MT) AND ("Chr"! = XY)].

(c) The above step will output only autosomal SNPs.

(d) Cluster separation: In the "SNP Table," sort by "Cluster Sep" in descending order (*see* **Note 23**).

(e) Rapidly scan through the SNPs to identify the rough lower limit for cluster separation. Zero all SNPs below this threshold. Thresholds used during our analysis were usually in the range (0.35–0.40). Sometimes, if the clusters are well defined, we can manually move or re-size the cluster position by clicking shift and moving the cursor (appears as a + sign) over the cluster. But as mentioned in the Illumina Technical Note, the choice to cluster manually should be considered very carefully. We recommend zeroing all abnormal clusters.

(f) Identifying low-intensity SNPs: In the "SNP Table," change Filter = [("Chr"! = X) AND ("Chr"! = Y) AND ("Chr"! = MT) AND ("Chr"! = XY) AND ("AB Freq"! = 0) and ("AB R Mean" < 0.2)] (*see* **Note 24**).

(g) The above step will output autosomal SNPs with low intensity; zero all of them.

(h) Identifying low-intensity SNPs: In the "SNP Table," change Filter = [("Chr"! = X) AND ("Chr"! = Y) AND ("Chr"! = MT) AND ("Chr"! = XY) AND ("AA Freq"! = 0) AND ("AA R Mean" < 0.2)].

(i) The above step will output autosomal SNPs with low intensity; zero all of them.

(j) Identifying low-intensity SNPs: In the "SNP Table," change Filter = [("Chr"! = X) AND ("Chr"! = Y) AND ("Chr"! = MT) AND ("Chr"! = XY) AND ("BB Freq"! = 0) AND ("BB R mean" < 0.2)].

(k) The above step will output autosomal SNPs with low intensity; zero all of them.

(l) Cluster separation based on AB T mean: In the "SNP Table," change Filter = [("Chr"! = X) AND ("Chr"! = Y) AND ("Chr"! = MT) AND ("Chr"! = XY) AND ("AB Freq"! = 0) AND ("AB T mean" < 0.2)] (*see* **Note 25**).

(m) The above step will output SNPs with AB clusters very close to the AA cluster. If feasible, the AB cluster can be manually re-clustered by pressing SHIFT + dragging the mouse. Otherwise, zero the SNPs.

(n) Unexpected number of samples in heterozygote (AB) cluster: In the "SNP Table," change Filter = [("Chr"! = X) AND ("Chr"! = Y) AND ("Chr"! = MT) AND ("Chr"! = XY) AND ("AB Freq" > 0.6)].

(o) The above step will output autosomal SNPs with AB Freq >0.6.

(p) Look for obvious multi-clusters in AB cluster. If these exist, they should be manually separated into different genotype clusters if possible or else zeroed. If there is a single well-clustered AB, then retain the SNP.

(q) In the "SNP Table," change Filter = [("Chr"! = X) AND ("Chr"! = Y) AND ("Chr"! = MT) AND ("Chr"! = XY) AND ("Het Excess" > 0.3)].

(r) The above step will output autosomal SNPs with Het excess >0.3; this identifies excess heterozygote calls relative to expected based on the Hardy-Weinberg equilibrium.

(s) Look for obvious multi-clusters in AB cluster. Follow as in step p.

(t) In the "SNP Table," change Filter = [("Chr"! = X) AND ("Chr"! = Y) AND ("Chr!= MT) AND ("Chr"! = XY) AND ("Het Excess" < −0.3)].

(u) The above step will identify SNPs with deficient heterozygote calls relative to expected based on the Hardy-Weinberg equilibrium. Zero these SNPs.

10. Final filtering.
    (a) From the "SNP Table" toolbar, select "Clear filter."
    (b) In the "SNP Table," change Filter = [("Call Freq" < 1) AND ("Chr"! = Y)] (*see* **Note 26**).
    (c) Zero all the SNPs from the above step.
11. Update SNP and sample statistics.
    (a) From the "SNP Table" toolbar, select "Clear filter."
    (b) In the SNP Table, "Select all" SNPs → right-click → click "Update Selected SNP Statistics."
    (c) In the Samples Table, "Select all samples" → right-click → click "Recalculate Statistics for all Samples" (*see* **Note 27**).
    (d) Repeat steps vi(a) but with a higher call rate threshold, e.g., 0.99 (*see* **Note 28**).
    (e) After the selected samples are excluded, you will be prompted to update sample and SNP statistics; click "Yes."
    (f) Click Save project (*see* **Note 29**).
12. Exporting data to PLINK format.
    (a) Download the plink plugin from Illumina website http://support.illumina.com/array/array_software/genomestudio/downloads.html/.
    (b) Analysis → Reports → Report Wizard → Custom Report → Plink Input Report 2.1.4 by Illumina → Report Input Parameters → UseForwardStrand (*see* **Note 30**) → True → Remove zeroed SNPs → Choose the folder and save the file.
    (c) Transfer *.ped, *.map, and phenotype files from the Windows workstation to Linux workstation using a secure FTP, such as MobaXterm.

### 3.10 Quality Control

This is one of the most important steps for a successful GWAS. Subtle biases in data can lead to spurious associations; thus, rigorous quality control steps should be implemented at every step [48]. The key concept is to detect poor-quality SNPs and/or samples prior to the association analyses. The steps above go some way to achieving this based largely on the signal intensity from the arrays, but here we outline methods for more refined quality control using standard genetic analysis tools. While there is no definite sequence as to which should be performed first, the steps are a trade-off between attempts to minimize type 1 error and loss of samples. Various metrics are used, and there is some level of subjectivity in the thresholds chosen for each step, depending on the study design, the sample size, and the sample cohort. Hundreds of packages are available for GWAS, with numerous tutorials

**Table 3**
**Files required for GWAS analysis in PLINK 1.9/2.0**

| File | Description |
|---|---|
| *.ped (e.g., DR.ped) | White space (space or tab)-delimited file with information on the individuals and their genotypes and phenotype |
| *.map (e.g., DR.map) | White space (space or tab)-delimited with information on the genetic markers (chromosomal positions) |
| Phenotype file (e.g., phenotype.txt) | Phenotype file with the details of all the phenotypes and covariates (*see* **Note 31**) |
| LD file (e.g., High_LD.txt) | Text file with list of SNPs in LD in hg19/GRCH37 and hg38/GRCH38 downloaded from https://genome.sph.umich.edu/wiki/Regions_of_high_linkage_disequilibrium_(LD) (for this protocol, we are using hg19/GRCH37) |

"*" should be replaced by the name of the file. In the examples below, we use "DR" as the generic filename for the main data held in DR.ped and DR.map files.

available online [12, 44, 49, 50]. For this chapter, QC procedures and statistical analyses will be illustrated using the free, open-source whole-genome association analysis toolset PLINK 1.9/2.0 [8, 9]. We use both versions interchangeably as PLINK 2.0 (https://www.cog-genomics.org/plink/2.0/) has some additional functions not present in PLINK 1.9 (https://www.cog-genomics.org/plink/) and vice versa. Excellent, detailed documentation on PLINK is available on the PLINK website. The files required for a standard QC run in PLINK are given in Table 3. For ease of understanding, we have used the example file names given in Table 3 for our QC steps described in this section and the analysis beyond.

### 3.11 QC Steps in PLINK

File names in the steps below correspond to file names given in Table 3. We begin with DR.ped and DR.map and the outputs after each stage with increasing integers, e.g., DR1, DR2, etc.

1. Change directory to a folder on your UNIX device containing all files for analysis in Table 3.
2. First convert the ped/map files to bed files.
   (a) plink --file DR --make-bed --out DR1
3. Investigate missingness per individual and per SNP.
   (a) plink --bfile DR1 –missing.
   (b) Output: plink.imiss shows proportion of missing SNPs per individual, and plink.lmiss shows the proportion of missing individuals per SNP.
   (c) Generate plots to visualize the missingness results (**Rcode1_miss**: Fig. 1).

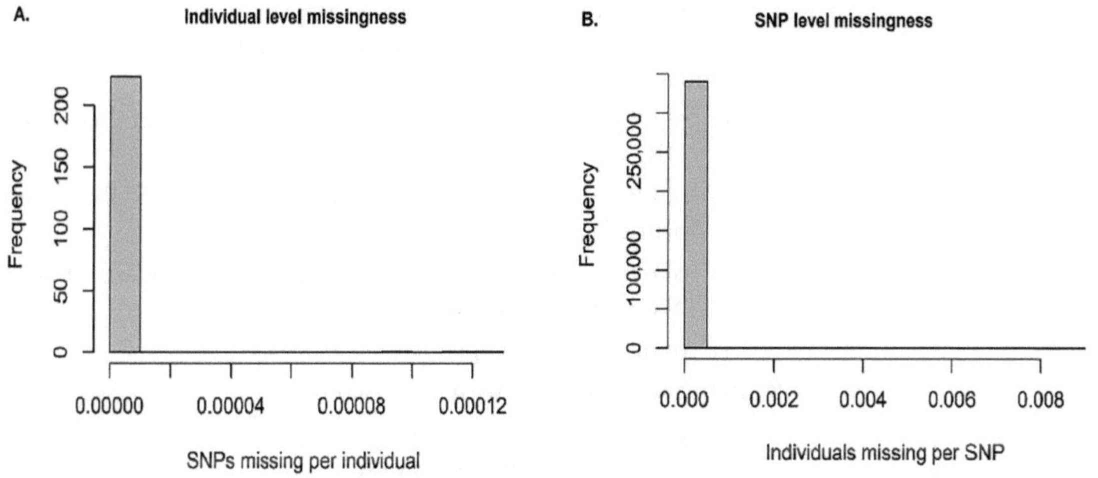


**Fig. 1** Distribution of missing data. Both the individual-level and SNP-level missingness in negligible in this example

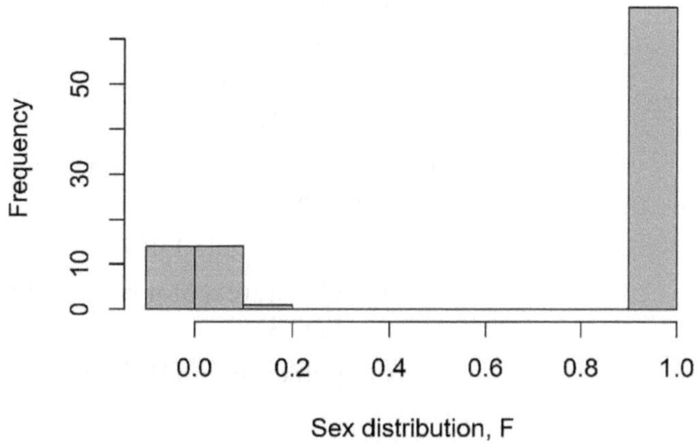


**Fig. 2** Plot showing sex distribution. There is a clear gap between a very tight male clump at the right-hand side of the distribution and the females on the left-hand side

4. Delete SNPs with high levels of missingness (*see* **Note 32**).
   (a) plink --bfile DR1 --geno 0.03 --make-bed --out DR2 (*see* **Note 33**)
5. Delete individuals with high levels of missingness.
   (a) plink --bfile DR2 --mind 0.03 --make-bed --out DR3
6. Checking for sex discrepancy (*see* **Note 34**).
   (a) plink --bfile DR3 --check-sex 0.40 0.80 (*see* **Note 35**)
   (b) Generate plots to visualize the sex check results (**Rcode2_sexcheck**: Fig. 2).

7. Pseudo-autosomal regions can sometimes give problematic sex check results. To avoid this, re-do sex check after excluding the pseudo-autosomal regions as below.
   (a) plink --bfile DR3 --split-x 'b37' no-fail --make-bed --out DR3sex
   (b) plink --bfile DR3sex --indep-pairwise 50 5 0.2 --out pruned
   (c) plink --bfile DR3sex --extract pruned.prune.in --make-bed --out DR3pruned
   (d) plink --bfile DR3pruned --check-sex
8. Remove individuals with sex discrepancy.
   (a) grep "PROBLEM" plink.sexcheck| awk '[10]' > sex_discrepancy.txt
   (b) plink --bfile DR3 --remove sex_discrepancy.txt --make-bed --out DR4
9. Selecting only autosomal SNPs.
   (a) awk '{if ($1 >= 1 && $1 <= 22) print $2}' DR4.bim > snp_1_22.txt
   (b) plink --bfile DR4 --extract snp_1_22.txt --make-bed --out DR5
10. Filtering SNPs with low minor allele frequency (MAF) (*see* **Note 36**).
   (a) plink --bfile DR5 –freq.
   (b) Generate a plot of the MAF distribution (**Rcode3_MAF**: Fig. 3).
   (c) plink --bfile DR5 --maf 0.01 --make-bed --out DR6.

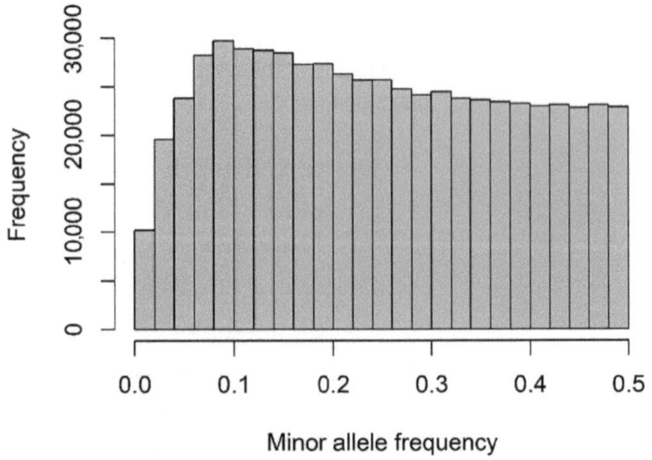


**Fig. 3** Minor allele frequency distribution. The minor allele frequency in this example is >0.05 for most of the SNPs

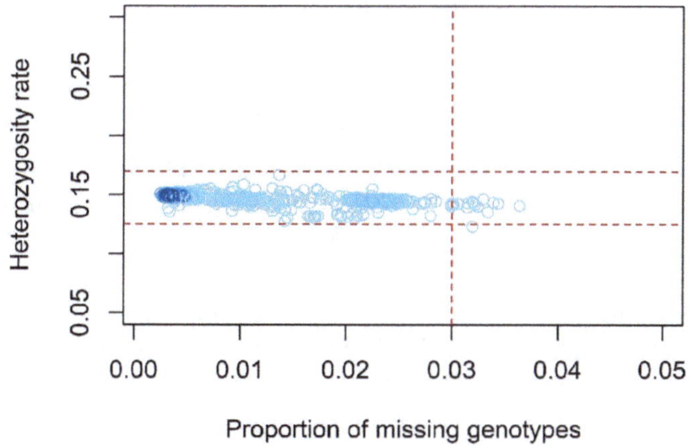


**Fig. 4** Plot of heterozygosity versus missingness. The blue dots represent individuals; dotted red lines indicate quality control thresholds. Based on this plot, we exclude individuals with missing genotypes >0.03 (vertical line) and individuals with heterozygosity > ±3 standard deviations from the mean (horizontal lines)

11. Hardy-Weinberg equilibrium filtering (*see* **Note 37**).
    (a) plink --bfile DR6 --hwe 1e-6 --make-bed --out DR7
12. Checking for heterozygosity.
    (a) plink --bfile DR7 --exclude High_LD.txt --range --indep-pairwise 50 5 0.2 --out indepSNP (*see* **Note 38**).
    (b) plink --bfile DR7 --extract indepSNP.prune.in --het --out Het_check.
    (c) Output: Het_check.het (# *the third column [O(Hom)] denotes the observed number of homozygous genotypes, and the fifth column [N(NM)] denotes the number of non-missing genotypes per individual*).
    (d) Then plot heterozygosity vs. missingness, and decide the cutoffs (usually ±3SD for heterozygosity) to exclude individuals (**Rcode4_HetMiss**: Fig. 4).
13. Removing individuals deviating > ± 3SD from mean heterozygosity or with excessive missing genotypes (*see* **Note 39**).
    (a) Add the family ID (FID) and individual ID (IID) of the samples failing this QC to the file "failqc.imisshet.txt."
    (b) plink --bfile DR7 --remove failqc.imisshet.txt --make-bed --out DR8.
14. Checking for cryptic relatedness.
    (a) plink --bfile DR8 --extract indepSNP.prune.in --genome --min 0.2 --out cryptic.
    (b) Output: "cryptic.genome" contains pairs with pi-hat value >0.2.

15. Deleting one individual per related pair with pi-hat >0.2 (# *remove the one with lower call rate*).
    (a) For each pair in "cryptic.genome," add the FID and IID of the individual with lower call rate to "low_call.txt."
    (b) plink --bfile DR8 --remove low_call.txt --make-bed --out DR9 (# *this file to be used for principal component analysis below*).

### *3.12 Principal Component Analysis to Control for Population Stratification*

Controlling for population stratification is an integral part of QC in GWAS [51]. One strategy for avoiding bias induced by population stratification is to ensure that study samples come from the same, relatively homogenous population. Principal component analysis (PCA) is now routinely used to detect and correct for population stratification in GWAS, even when all participants report the same background ethnicity [52]. By anchoring the study dataset against a reference genome of known ancestry, outliers likely to have different ancestry are excluded. Here, we use 1000 Genomes dataset, phase 3 Ver. 5 [53] (1000G) as a reference population containing individuals of European (EUR), Asian (ASN), admixed American (AMR), and African (AFR) ancestry. The basic steps for PCA are given below. This step is conducted by merging the cleaned study dataset (DR9 from above QC) to 1000G data from the four populations.

1. First download the 1000G data from the official plink webpage https://www.cog-genomics.org/plink/2.0/resources (hg37 build).
    (a) Click the three blue bold faced links, "**pedigree corrected**," "**.pgen.zst**," and "**.pvar.zst**," to download the respective files.
2. <OR> download the three files from Dropbox as below.
    (a) wget https://www.dropbox.com/s/afvvfl e15gqzsqo/all_phase3.pgen.zst?dl=1
    (b) mv 'all_phase3.pgen.zst?dl = 1' all_phase3.pgen.zst (# *changing name to a more suitable form using mv*)
    (c) plink2 --zst-decompress all_phase3.pgen.zst > all_phase3.pgen (# *decompressing the file*; *see* **Note 40**)
    (d) wget https://www.dropbox.com/s/op9osq6luy3pjg8/all_phase3.pvar.zst?dl=1
    (e) mv 'all_phase3.pvar.zst?dl = 1' all_phase3.pvar.zst
    (f) plink2 --zst-decompress all_phase3.pvar.zst > all_phase3.pvar (# *decompressing the file*)
    (g) wget https://www.dropbox.com/s/yozrzsdrwqej63q/phase3_corrected.psam?dl=1
    (h) mv 'phase3_corrected.psam?dl = 1' all_phase3.psam

3. Convert the above data from *.pgen format to PLINK 1.9 binary format (*see* **Note 41**).
   (a) plink2 --pfile all_phase3 --max-alleles 2 --make-bed --out All_phase3 (*see* **Note 42**).
4. It is recommended to perform a baseline QC of the 1000G dataset (shown below) for the purpose of merging with the study dataset (*see* **Note 43**).
   (a) plink --bfile All_phase3 --geno 0.02 --mind 0.02 --maf 0.05 --make-bed --out All_phase3a
   (b) plink2 --bfile All_phase3a --set-all-var-ids @:#[b37]$r,$a --rm-dup force-first --make-bed --out All_phase3b (# *using PLINK 2 for assigning unique IDs*)
   (c) plink2 --bfile All_phase3b --snps-only just-acgt --make-bed --out GENOME1KG (# *this file will be used for merging below*)
5. Next merge the study dataset (DR9) with 1000G dataset (GENOME1KG).
   (a) Both datasets must be in the same genome build (use LiftOver [54], http://genome.ucsc.edu/cgi-bin/hgLiftOver, to change genome build in case of difference in genome build).
   (b) Both datasets must have the same SNP ID format (shown below).
   (c) plink2 --bfile DR9 --set-all-var-ids @:#[b37]$r,$a --rm-dup force-first --make-bed --out DR10 (# *changing to similar ID format as GENOME1KG*; *see* **Note 44**).
   (d) plink --bfile DR10 --bmerge GENOME1KG --make-bed --out MERGED (*see* **Note 45**).
6. Running PCA on the merged dataset (first prune the dataset; then run PCA).
   (a) plink --bfile MERGED --exclude High_LD.txt --range --indep-pairwise 50 5 0.2 --out Indep
   (b) plink --bfile MERGED --extract Indep.prune.in --pca --out mergePCA
   (c) Output: mergePCA.eigenvec (# *used for PCA plotting*) and mergePCA.eigenval.
7. Plotting the principal coordinates. For this, we need population information for the 1000G samples. The "all_phase3.psam file" downloaded above in **step 2** has superpopulation and population information for each sample in the 1000G dataset.
   (a) awk '{print $1, $5}' all_phase3.psam > superpop_1kG.txt (# *ancestry file for reference dataset*)

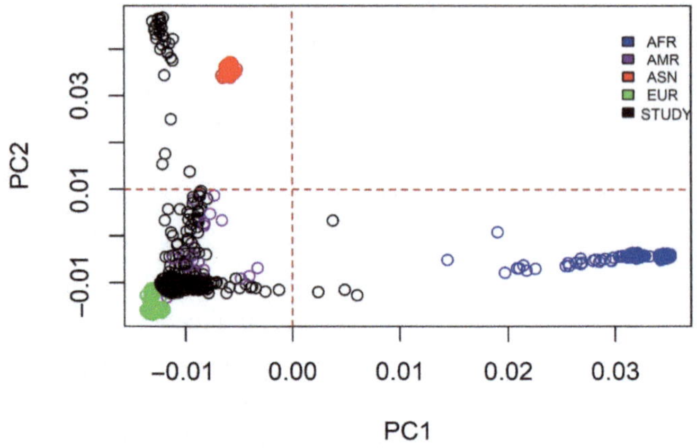


**Fig. 5** Plot of principal component 1 (PC1) versus principal component 2 (PC2). Based on this plot, we exclude individuals with PC1 > 0.00 (vertical line) or PC2 > 0.01 (horizontal line)

(b) awk '{print $1, $2, "STUDY"}' DR9.fam > ancestryfile_STUDY.txt (*# each individual in the study dataset is assigned ancestry "STUDY"*)

(c) cat superpop_1kG.txt ancestryfile_STUDY.txt | sed -e '1i \FID IID ancestry' > ancestry_SUPERIMPOSED.txt.

(*# creating a common ancestry file for the merged dataset, this file to be used for PCA plotting*)

Plot PCA bi-plot in R (**Rcode5_PCA.R**: Fig. 5).

8. Examine the PCA plot and add the FID and IID of the ancestry outliers to the file Outlier.txt.
9. Remove the ancestry outliers from the study dataset.
   (a) plink --bfile DR9 --remove Outlier.txt --make-bed --out CleanData
   (b) Output: CleanData.bed/CleanData.bim/CleanDate.fam (*# this file to be used for imputation*).
10. Next, perform PCA only on the clean study dataset. This is to generate the principal components to be used as covariates in the association analysis later.
    (a) plink --bfile CleanData --extract indepSNP.prune.in --pca --out covarPCA.
11. Extract the first ten principal components to use as covariates.
    (a) awk '{print $1, $2, $3, $4, $5, $6, $7, $8, $9, $10, $11, $12}' covarPCA.eigenvec > covarPC.txt.
12. Join the principal components to the phenotype file.
    (a) Join Phenotype.txt covarPC.txt > PhenotypePC.txt (*# this file to be used as phenotype/covariate file for association analysis*).

### 3.13 Data Imputation

Genotype imputation is a key component of genetic association studies that can increase their resolution by inferring genotypes at untyped variants, increasing the overall number of markers [55]. In the following section, we describe the imputation process using the Michigan Imputation Server [10] https://imputationserver.sph.umich.edu/index.html#. We recommend a thorough QC of the study dataset before imputation. The steps described below and the files used are adapted from the guidelines developed by the Michigan Imputation Server Team (https://imputationserver.readthedocs.io/en/latest/prepare-your-data/).

### 3.14 Files Required for Imputation

1. The cleaned plink file from above: CleanData.bed/CleanData.bim/CleanData.fam.
2. wget http://www.well.ox.ac.uk/~wrayner/tools/HRC-1000G-check-bim-v4.2.7.zip (# *from McCarthy Group Tools (Will Rayner)*) (*see* **Note 46**).
3. wget ftp://ngs.sanger.ac.uk/production/hrc/HRC.r1-1/HRC.r1-1.GRCh37.wgs.mac5.sites.tab.gz (# *from McCarthy Group Tools (Will Rayner)*).
4. wget http://qbrc.swmed.edu/zhanxw/software/checkVCF/checkVCF-20140116.tar.gz (# *from gitHUB repository*).
5. Unzip the checkVCF file. It has four files—checkVCF.py, two FASTA files, and an example.vcf.gz file.

### 3.15 Data Preparation

1. The data to impute will be the cleaned data post QC and PCA described above in binary plink format (CleanData.bed/CleanData.bim/CleanData.fam).
2. Create a frequency file.
   (a) plink --freq --bfile CleanData --out freqCleanData
3. Run the recommended McCarthy Group (Will Rayner)-authored scripts downloaded above. The scripts check a cleaned plink *.bim file against the Haplotype Reference Consortium (HRC) or 1000G reference SNP list in advance of imputation.
   (a) perl HRC-1000G-check-bim.pl -b CleanData.bim -f freqCleanData.frq -r HRC.r1-.GRCh37.wgs.mac5.sites.tab -h (*see* **Note 47**) (# *this will output a script file "Run-plink.sh"*)
   (b) sh Run-plink.sh (*see* **Note 48**) (# *this will output vcf files for each chromosome along with individual corrected CleanData-updated-chr*.bed/bim/fam files*)
4. Create a sorted vcf.gz file for each chromosome using BCFtools.
   (a) bcftools sort CleanData-updated-chr*.vcf -Oz -o Chr*.vcf.gz

5. Check *.vcf files are valid for each chromosome.
   (a) python checkVCF.py -r hs37d5.fa -o test Chr*.vcf.gz (*see* **Note 49**).
   (b) Error report will be generated in case of invalid VCF files.

### *3.16 Upload the Project to the Michigan Imputation Server*

1. Upload the files (vcf file per chromosome) to the Imputation Server (https://imputationserver.readthedocs.io/en/latest/getting-started/#setup-your-first-imputation-job).
2. Choose the most appropriate reference panel as per the study cohort.

### *3.17 Post-imputation Processing*

1. Download the file from the server using the password emailed to you.
   (a) unzip -P password '*.zip'
2. Extract good-quality imputed genotype data in dosage format (*# PLINK 2 is used for this because PLINK 1.9 does not handle dosage format*).
   (a) plink2 --vcf *.vcf.gz dosage = DS --exclude-if-info "R2 < 0.8" --make-pgen --out ImputedDR (*see* **Note 50**)
3. We recommend another round of QC post-imputation as described in the "Quality control section." QC in PLINK 2 for dosage data uses *.pgen files, so use "--pfile" and "--make-pgen" in place of "--bfile" and "--make-bed," respectively.
4. Output: ImputedDRclean.pgen/ImputedDRclean.pvar/ImputedDRclean.psam (*# to be used for association analysis*).
5. Alternatively, one could generate *.bim/*.bed/*.fam files as below and then continue with the QC steps and the rest of the analysis in either PLINK 1.9 or PLINK 2.0. However, this format does not include dosage information, so we prefer using the dosage format in PLINK 2.0.
   (a) plink2 –vcf *.vcf.gz dosage = DS --exclude-if-info "R2 < 0.8" --make-bed --out ImputedDR.

### *3.18 Association Analysis (Using the Dosage Format in PLINK 2.0)*

The main goal of GWAS is to identify variants associated with a phenotype of interest. In general, it is assumed that these associated variants are not the causative variants but are a proxy in linkage disequilibrium (LD) with the causative variants. Post-GWAS analysis, replication, in vitro experimentation, and prospective studies are usually required to infer causality. Various association models can be applied for DR-related GWAS depending on the phenotype. Logistic and linear regression under additive models (shown below) adjusting for necessary confounding factors are the two most commonly used approaches for case-control and quantitative phenotypes, respectively. Population stratification is a common cause for spurious association in GWAS; therefore, it is pertinent to include

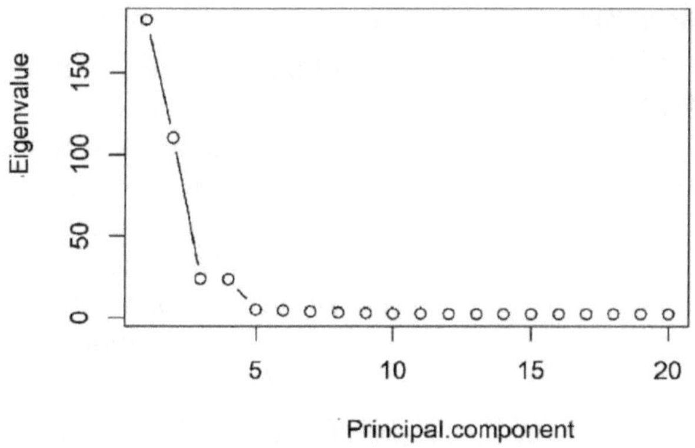


**Fig. 6** Scree plot showing the eigenvalues of the first 20 principal components (PCs). This plot depicts an elbow at the third PC, indicating that the first three PCs account for a significant proportion of the population structure and should be included as covariates in the analysis

principal components (PCs) to account for stratification [52]. Though there are no standard ways to decide on the number of PCs to include, inspection of PCA scatter plots to determine which PCs are best at separating the cohort population can be informative. For example, in Fig. 5, the samples are separated in the first two PCs so we include PC1 and PC2 as covariates as a minimum. Another way is to visualize the "Scree plot for elbow test" and choose the PCs to account for 80 ~ 90% of the variation explained [56]. The scree plot involves plotting the eigenvalues (y-axis) against the principal components (x-axis) and choosing a point in the plot where there is a sharp change in the slope (**Rcode6_ScreePlot**: Fig. 6). This point on the plot indicates the maximum number of PCs to retain [57]. Including more PCs can help to adjust for more subtle differences between the samples; however, it is important to avoid overfitting of the data and smaller studies should consider only including significant principal components. The degree of remaining stratification can be indicated by the genomic inflation factor ($\lambda$). This should be close to 1.00 to indicate control of population stratification [58]. For quantitative traits, transformation of data is recommended in the event of non-normal distribution of errors [59].

**Files Required**

1. Post-imputation cleaned files from above: ImputedDRclean.pgen/ImputedDRclean.pvar/ImputedDRclean.psam.
2. Covariate file with details of all the covariates, phenotypes, and PCs to be used: PhenotypePC.txt (*see* **Note 51**).

### 3.19 Association Steps for Different Study Designs

1. Case-control design (logistic regression).
   (a) plink2 --pfile ImputedDRclean --glm hide-covar cols = +a1freqcc --ci 0.95 --pheno PhenotypePC.txt --pheno-name BinaryPheno --covar PhenotypePC.txt --covar-name covar1, covar2, PC(1–2) --covar-variance-standardize --adjust --out DR_CC (*see* **Note 52**) (# *"BinaryPheno here" is the name of the phenotype as listed in the column header, and "covar1, covar1, PC (1-2)" are the covariate column headers in the PhenotypePC.txt phenotype file*).
   (b) Output: DR_CC.glm.logistic (*see* **Note 53**) (# *to be used for Manhattan and QQ plot*) and DR_CC.glm.logistic.adjusted.
   (c) The *.adjusted file has a Bonferroni corrected p-value, along with false discovery rate (FDR) and other statistics.
   (d) gzip DR_CC.glm.logistic.
   (e) Output: DR_CC.glm.logistic.gz (# *to be used for regional plot*).
2. Quantitative variable (linear regression).
   (a) plink2 --pfile ImputedDRclean --glm hide-covar --ci 0.95 --pheno PhenotypePC.txt --pheno-name LinearPheno --covar PhenotypePC.txt --covar-name covar1, covar2, PC (1–2) --covar-variance-standardize --adjust --out DR_Linear,
   (b) Output: DR_Linear.glm.linear (to be used for plotting) and DR_Linear.glm.linear.adjusted.
   (c) gzip DR_Linear.glm.linear.
   (d) Output: DR_Linear.glm.linear.gz (# *to be used for regional plot*).

### 3.20 Generating Manhattan and QQ Plots

This can be done in the Linux environment or separately in R/RStudio. The code is provided in Rcode7_manQQ: Fig. 7.

***Files required*** (summary-level data: chromosome, position, rsID, *p*-value)

1. DR_CC.glm.logistic (for logistic regression).
2. DR_Linear.glm.linear (for linear regression).

### 3.21 Generating Regional Plots

There are several software packages and web tools for plotting and visualizing regional plots of SNPs of interest [13, 60, 61]. The web portal FUMA [14] is an easy-to-use option. Readers are referred to the article by Watanabe et al. [14] and a well-illustrated tutorial posted on the official FUMA website (https://fuma.ctglab.nl/tutorial) for a more comprehensive understanding of the steps (*see* **Note 54**).

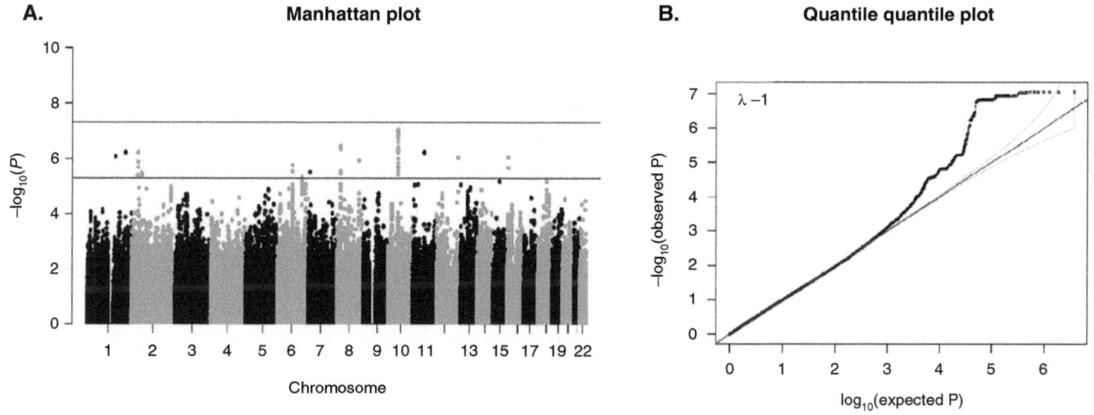


**Fig. 7** Plots for visualizing genome-wide association study results. (**a**) Manhattan plot showing negative log of the *p*-value for the association statistic (*y*-axis) for each SNP plotted against its position in the genome (*x*-axis). The red and blue horizontal lines represent the genome-wide significance threshold of $p < 5.0 \times 10^{-8}$ and threshold for suggestive significance of $p < 5 \times 10^{-5}$, respectively. (**b**) Q-Q (quantile-quantile) plot showing the expected –log10 *p*-value under the null hypothesis on the *x*-axis and the observed on the *y*-axis, adjusting for covariates. The red line represents $x = y$ and the gray area represents the 95% confidence interval. The lambda value ($\lambda$) is a measure of the genomic inflation factor and should be close to 1.00

***Files required*** (summary-level data: chromosome, position, rsID, p-value)

1. DR_CC.glm.logistic.gz (for logistic regression).
2. DR_Linear.glm.linear (for linear regression).

### 3.22 Validation and Replication Analysis

It is always prudent to validate the results of the association test. Most DR studies with significant association suffer from lack of replication in an independent cohort [7, 35]. Considering the fact that DR is a complex and heterogeneous phenotype with relatively moderate to small effect size for genetic risk variants, the sample size of the replication cohort should be larger than the discovery cohort, particularly to account for the well-described phenomenon of winner's curse [62] where the chance finding of a larger effect size results in publication. Like all other complex diseases, it is extremely important for the replication cohort to have the same phenotype definition as the original discovery cohort. Since the phenotypic heterogeneity in DR is wide, using even slightly different phenotypic definitions can adversely affect the outcomes of the replication analysis if the genetic risk is specific to a particular subtype or feature. It is also important to adjust for the same covariates as the original cohort as well as any additional confounding factors that might have been identified in the replication cohort. Another method for validation would be a meta-analysis approach taking multiple DR-related GWAS studies of identical (or as similar as possible) design and phenotype. For meta-analysis in general, it is rare to find multiple studies that match perfectly on all criteria, and considering the inherent nature of DR heterogeneity, one can expect the study-level heterogeneity to be prominent. Use of an extreme DR phenotype might help to reduce such bias. Finally, a thorough and detailed QC process is one of the basic and most important requirements for a successful GWAS. Compromised data quality in the discovery phase may lead to false-positive results that are carried forward into replication studies at great cost in both time and expense.

### 3.23 Future Directions

One of the first and most important bottlenecks commonly encountered when working with DR-related traits is getting a homogenous phenotype and a true-control cohort in adequate numbers. The DR cases studied to date tend to be of varying levels of severity and disease definitions, thus increasing the heterogeneity. Likewise, the controls/pseudo-controls in DR usually have good diabetes control and shorter duration of diabetes and consequently have a strong possibility of developing DR in the future. An ideal solution would be to establish large collaborative prospective cohorts to study diabetes-related complications including DR, but this is logistically challenging. Data from established or ongoing data repositories such as the UK Biobank (https://www.ukbiobank.ac.uk) will provide opportunities for well-powered studies across many phenotypes including DR [63, 64]. Artificial intelligence and machine learning approaches are also being applied to GWAS for complex disorders to enhance GWAS performance [65, 66]. For instance, machine learning has been applied to identify loci, increase the statistical power of GWAS [67], detect epistatic interactions [68], and improve polygenic risk scoring produced from GWAS [69] and for post-GWAS prioritization [70]. Using a similar approach, machine learning combined with GWAS in DR could achieve increased predictive performance of DR risk.

## 4 Notes

1. Ensure that the initial data entry is done very thoroughly using well-trained staff. Make sure to perform appropriate data validation.
2. Use of standard conversion techniques from Snellen's visual acuity to ETDRS or logMAR scale is recommended.
3. There is an increased measurement in retinal thickness of approximately 65 to 70 microns, as measured by Spectralis optical coherence tomography (Spectralis OCT) compared with Stratus OCT, corresponding to the inclusion of the outer segment-RPE-Bruch's membrane complex by the Spectralis OCT; hence, care should be taken to correct for this discrepancy between the two OCT devices while recording data from medical records [71].

4. For pre-study power calculation, we calculate power for a range of allele frequencies from 0.01 to 0.5.
5. The relative risk is the odds ratio for risk allele carriers compared to non-carriers. This is a predicted value. We recommend exploring a range of values (e.g., 1.25, 1.50, 1.75, 2.0).
6. This is the odds ratio for people carrying two risk alleles. It should be set based on the model to be used (additive, dominant, multiplicative, etc.).
7. Marker allele frequency is the frequency of the genotyped allele if that is not expected to be the actual causative variant. If using 1 for LD, we set this to the same as the high-risk allele frequency.
8. Unselected controls should be checked if the controls have not been screened for the disease, i.e., they are general population controls.
9. Only one vial is used for DNA extraction. It is optimal to collect samples in duplicate to provide an extra in case of future accidents in transport or handling or if researchers need to validate results later.
10. All these files are provided by the genotyping facility or downloadable from the product-specific support page on Illumina.com (e.g., for GSA array: https://sapac.support.illumina.com/array/array_kits/infinium-global-screening-array/downloads.html). The files are specific to a particular array design/version and to the genome build. The name of the manifest file in the sample sheet MUST match the name of the manifest file being used.
11. Templates for Sample Sheet are available under Support Files on Illumina.com. Sample Sheet must be in the .csv format with at least the first three columns comprising Sample_ID, SentrixBarcode_A, and SentrixPosition_A. The Manifest name must exactly match the .bpm file being used. Ensure that the SentrixBarcodes_A are not converted to scientific notation (format the column as "Numbers"). Additional information such as age and sex can be added to this file before loading the project. The SentrixBarcode_A column should be formatted as numbers every time we edit the Sample Sheet. The file should be closed before we load the project.
12. IDAT files by convention are processed using GenomeStudio to generate a plain-text ASCII file for further downstream analysis. There are other alternative parsing tools available, e.g., *illuminaio* [72], although they have a greater need for extensive bioinformatics skills. Therefore, we recommend using GenomeStudio.

13. Make sure all the files are in a single repository folder for loading the project in GenomeStudio.
14. Projects can be created without the cluster file, but we recommend using the standard cluster file because the genetic diversity of the samples is well represented by the standard cluster file unless in instances when the call rate is low.
15. Depending on computer memory and the number of samples/SNPs, this may take several minutes to hours to cluster the data and calculate statistics. The no-call threshold of 0.15 ensures that genotypes with a GenCall score less than 0.15 are not assigned genotypes.
16. Always ensure that the sample numbers shown in GenomeStudio match those in the sample sheet. If there are any errors in the sample sheet or a corresponding .idat file is not present at the given location, GenomeStudio will not upload the sample in question and will not prompt an error.
17. It is recommended to save a raw copy of the original project if we need to repeat the QC steps in the future with a different set of parameters. Saving a separate raw copy of the project will save time in loading the project, especially for large datasets.
18. Some researchers consider the steps from iv to x optional; instead, they directly export the data into other suitable formats for QC in other software.
19. Call rate represents the percentage of all SNPs and indels that are successfully genotyped for a given sample. The threshold for call rate varies with study. Poor call rate might not be due to poor sample quality but due to biological causes such as tumor cells, and care should be taken when removing samples based on call rate alone. Start with a lower call threshold (e.g., 95%) for initial removal, and decide later if samples with intermediate call rates (95–99%) consistently show poor performance in other parameters as well.
20. It is recommended to update the sample and SNP statistics at the end just before the final filtering step. Updating statistics is time-consuming.
21. No "gender est." is assigned in GenomeStudio for samples with a call rate < 0.95. In case of a mismatch between "gender" (reported sex) and "gender est." (sex predicted by software based on genetic data), it is recommended to note the sample ID rather than removing them. If, while checking sex discordance in PLINK, the same samples show problems later on, then exclude the samples or re-process them or check for any data entry error.

22. The number of SNPs to be manually checked in this step is arbitrary. We recommend reviewing all of the X chromosome SNPs with an ABFreq >0, though some recommend checking at least the first 100–200 SNPs.
23. Cluster separation measures the separation between the three genotype clusters and varies from 0 to 1. The clusters should be approximately equally spaced. Poorly clustered genotypes should be zeroed.
24. AB R mean represents the mean normalized intensity (R) of the heterozygote cluster. This metric helps identify SNPs with low-intensity data and has values increasing from 0.
25. AB T mean is the mean theta intensity (T) of the heterozygote cluster. The optimal value for AB T mean should be 0.5. If the value is shifted toward 0 or 1, it means that there is a higher likelihood of erroneous genotype.
26. Freq <1 can be altered to whatever threshold is appropriate for the study. Don't forget to exclude Y chromosome markers for this step.
27. For this step, we select all samples, including those we excluded in **step 6 (a)**. Filtering out the poorly performing SNPs often improves the final call rate of the samples.
28. The threshold for call rate is study dependent, and it is common to use 0.95; we generally prefer a higher threshold of $\geq$0.99.
29. GenomeStudio does not autosave, so it is highly recommended to click the Save button after every filtering step.
30. Illumina has its own strand orientation nomenclature (https://www.illumina.com/documents/products/technotes/technote_topbot.pdf), so it is necessary to be careful during the downstream analysis. An alternative way would be to transfer the file to the Illumina top-bottom format and convert it to the hg19 forward strand format as per the detailed instructions provided in https://www.well.ox.ac.uk/~wrayner/strand/. Whichever strand orientation is used, it must remain uniform throughout the analysis process.
31. A single phenotype can be encoded directly in the *.fam file or *.ped file, but we recommend using a common *.txt file for all the phenotypes and covariates (including the principal component). This approach allows simple specification of the phenotypes and covariates for each analysis as required without extensive file reformatting.
32. Whether to perform individual-level QC first or SNP-level QC first has different pros and cons [73, 74]. The choice should be guided by study cohort, sample size, and study design.

33. Missingness can be due to poor DNA quality, genotyping platform, or technical errors, and the exact threshold varies from study to study. The threshold should be determined to minimize the number of samples dropped while maximizing genotyping efficiency. Typically, we use a threshold between 2% and 5%.
34. Sex is of relevance only when carrying out a sex-stratified analysis or when analyzing the X chromosome. However, a sample with discordant sex information might indicate sample handling errors and a possible incorrect phenotype-genotype combination.
35. The default F value <0.2 for female and > 0.8 for male thresholds are not applicable for every dataset. We generally run "--check-sex" once without parameters and plot the sex distribution (as recommended in the PLINK documentation). Ideally, there should be a clear gap between a very tight male clump at the right side of the distribution and the females everywhere else (Fig. 2). Then we re-run with parameters corresponding to the empirical gap. If any samples remain problematic, wherever feasible, always double-check the data entry has been done correctly and the correct sex has been assigned.
36. MAF threshold depends strongly on sample size and the expected effect size. Most studies set the MAF threshold in the range of 0.01 to 0.05 to avoid analyzing SNPs that are clearly underpowered.
37. Though HWE can indicate potential genotyping error, it can also result from a true genetic association with disease risk [75]. We recommend performing HWE only for controls as disease-free controls should more closely follow the assumptions that lead to HWE than cases. By default, PLINK only filters HWE for controls in a case-control design.
38. Heterozygosity should be calculated for a set of SNPs that are not highly correlated, so we exclude SNPs in linkage disequilibrium (LD), described in Table 3.
39. High levels of heterozygosity may indicate poor sample quality or contamination, whereas low levels of heterozygosity may be due to inbreeding. We decide on exclusion cutoffs so as to identify outliers by looking at the plot. We generally use ±3 SD, but when we suspect mixed ancestry or ± 3SD removes too many individuals, then we use a less stringent QC cutoff {± (4–6) SD} [74] based on the plots generated from "Rcode4_HetMiss."
40. PLINK 2 is used for *.pgen files, and also it can directly decompress .zst files.

41. Converting to PLINK 1.9 format because merging is easier in PLINK 1.9 than in PLINK 2.
42. *.bim cannot contain multiallelic variants, and PLINK returns errors. Use --max allele 2 to filter to only biallelic variants.
43. QC will remove indels, duplicates, and non-ACGT markers, which may cause problems with merging in PLINK. You can merge using other tools as well (e.g., BCFtools).
44. This step is in PLINK2 and will avoid errors caused by multiple variants with the same position during the merge in PLINK 1.9.
45. Strand issues may give merge errors, in which case perform trial flip as explained in detail in https://www.cog-genomics.org/plink/1.9/data#merge3.
46. The details of the files and the steps for pre-imputation check are given at https://www.well.ox.ac.uk/~wrayner/tools/.
47. The script will check strand, alleles, position, Ref/Alt assignments, and frequency differences.
48. The script will update strand, position, and ref./alt assignment.
49. Ensure the genome build of the study dataset matches the FASTA file genome build.
50. For imputation quality, we generally use a stringent $R^2 > 0.8$ to filter for good-quality imputation, though one should be aware that for common alleles, one can choose a lower value, e.g., >0.3. $R^2$ is the measure of the correlation between imputed genotype calls and the true underlying genotypes, i.e., the ratio of the variances of imputed and true allele based on HWE counts. We can convert to PLINK 1.9 format using the "--make-bed" command.
51. One can include the phenotype in the *.psam file or *.fam file, but we prefer using a common file for all phenotypes and covariates, including principal components.
52. We use the option "hide-covar" to only show the additive model. This model has consistently good power over a wide range of genetic models and is the most commonly used in GWAS. Use "--covar-variance-standardize" because sometimes PLINK generates errors if the genotype/covariate scales vary too widely; "--adjust" will give results as per multiple test correction and also will generate genomic inflation factor ($\lambda$). As a correction for multiple testing, we typically set the significance level at $5 \times 10^{-8}$. This roughly corresponds to a simple Bonferroni correction to maintain a 5% genome-wide type 1 error rate that is based on an estimated burden of one million independent tests [76]. The allele frequency for cases and controls is given by "cols = +a1freqcc."

53. The *.logistic file output gives the association results per individual marker (SNP), including effect size (beta, odds ratio), standard error, and p-values. A detailed description of the output parameters is given on the official PLINK website.
54. Before uploading the association output summary statistics in the FUMA web portal, always ensure the column headings/ names are correct; remove any symbols, e.g., "#," from PLINK 2 output files.

## 5 R Codes for the Protocol to Run in RStudio/R

### 5.1 Rcode1_miss (Visualizing per Individual and per SNP Missingness, Fig. 1)

*Files required:* Output files from running "--missing" flag in quality control step.

(i) plink.imiss (proportion of missing SNPs per individual).

(ii) plink.lmiss (proportion of missing individuals per SNP).

Individual-level missingness

- `indmiss <- read.table(file = plink.imiss, header = TRUE)` *# read data in R*
- `png(Individual_missingness.png, units = in, width = 5, height = 4, res = 300)`

  *# denotes the image format and resolution*
- `hist(indmiss[,6], main = Individual level missingness, xlab = SNPs missing per individual)`

  *# column 6 has the values corresponding to proportion of SNPs missing per individual*
- `dev.off()` *# tells R that we are finished plotting*

SNP-level missingness

- `snpmiss <- read.table(file = plink.lmiss, header = TRUE)`
- `png(SNP_missingness.png, units = in, width = 5, height = 4, res = 300)`
- `hist(snpmiss[, 5], main = SNP level missingness, xlab = Individuals missing per SNP)` *# column 5 has the values corresponding to proportion of individuals missing per SNP*
- `dev.off()`

### 5.2 Rcode2-_sexcheck.R (Visualizing Sex Distribution, Fig. 2)

*Files required*: Output file from running "--sex-check" flag in quality control.

(i) plink.sexcheck.

Generating histogram

- `sex <- read.table(plink.sexcheck, header = TRUE)`

- `png(Sex_distribution.png, units = in, width = 5, height = 4, res = 300)`
- `hist(sex[, 6], main = Sex distribution, xlab = F)`

  *# column 6 has F values*
- `dev.off()`

### 5.3 Rcode3_MAF (Generating Plot for Minor Allele Frequency (MAF) Distribution, Fig. 3)

*Files required:* Output file from running "--freq flag" in quality control.

(i) plink.frq.

Generating histogram

- `maf <- read.table(plink.frq, header = TRUE)`
- `png(MAF distribution.png, units = in, width = 5, height = 4, res = 300)`
- `hist(maf[, 5], main = MAF distribution, xlab= MAF)`

  *# column 5 has the MAF values*
- `dev.off()`

### 5.4 Rcode4_HetMiss (Generating Plot for Heterozygosity vs. Missingness, Fig. 4)

*Files required:* Output files from running "--het" and "--missing" flags.

(i) Het_check.het.

(ii) plink.imiss (make sure both files have the same number of individuals. If individuals have been removed in previous steps, then the number of individuals will differ between the two files, so make sure to remove those individuals from the .imiss file).

Generating the plot

- `imiss <- read.table(plink.imiss, head = TRUE)`
- `het <- read.table(Het_check.het, head = TRUE)`
- `het$meanHet <- (het$N.NM.-het$O.HOM.)/het$N.NM.`
- `png(Het_Miss.png,units = in, width = 5, height = 4, res = 300)`
- `colors <- densCols(imiss$F_MISS,het$meanHet)` *# creates colored scatter plot*
- `plot(imiss$F_MISS, het$meanHet, col=colors, xlim = c(0,0.05), ylim = c(0.00,0.3), xlab = Proportion of missing genotypes, ylab = Heterozygosity rate)`
- `abline(h = mean(het$meanHet) - (3*sd(het$meanHet)), col = RED, lty = 2)` *# the standard deviation (sd) value can change depending on the dataset*
- `abline(h = mean(het$meanHet) + (3*sd(het$meanHet)), col = RED, lty = 2)`
- `abline(v = 0.03, col = RED, lty = 2)`
- `dev.off()`

### 5.5 Rcode5_PCA.R (Generating Plot for Principal Component Analysis to Check for Population Stratification, Fig. 5)

*Files required:* Output files from running "--pca" and the ancestry file generated from all_phase3.psam.

(i) mergePCA.eigenvec.

(ii) ancestry_SUPERIMPOSED.txt.

Generating the plot

- `install.packages(RColorBrewer)`
- `library(RColorBrewer)`
- `options(scipen = 100, digits = 3)`
- `eigenvec <- read.table(mergePCA.eigenvec, header = FALSE, skip = 0, sep = ')`

  *# loading the above .eigenvec file*
- `eigenvec$V1 <- eigenvec$V2`
- `colnames(eigenvec)[1] = Fam.ID`
- `colnames(eigenvec)[2] = Ind.ID`
- `ancestry <- read.table(file = ancestry_SUPERIMPOSED.txt, header = TRUE)` *# loading the above ancestry file*
- `colnames(ancestry)[1] = Fam.ID`
- `colnames(ancestry)[2] = Ind.ID`
- `PC <- merge(eigenvec, ancestry, by = c(Fam.ID, Ind.ID))`
- `rownames(PC) <- PC[, 2]`
- `PC <- PC[, 3:ncol(PC)]`
- `colnames(PC) <- paste(PC', c(1:20), sep = ')`
- `colnames(PC)[21] <- Ancestry`
- `png(Population Stratification.png, unit = in, width = 5, height = 4, res = 300)`

  `col <- colorRampPalette(c(blue, purple, red, green, black))(length(unique(PC$Ancestry)))[factor (PC$Ancestry)]`
- `plot(PC[, 1], PC[, 2], col = col, xlab = PC1, ylab = PC2)`
- `abline(v=0.00, col = RED, lty=2)`
- `abline(h=0.01, col = RED, lty=2)`
- `legend(topright, bty=n, cex=0.6, title=, c (AFR, AMR, ASN, EUR, STUDY), fill = c(blue, purple, red, green, black))`
- `dev.off()`

Detecting the outliers

- `outliers <- PC[which(PC$PC1 > 0 | PC$PC2 > 0.01), 1:2]`
- `write.table(outliers, "outliers.txt", quote = F, sep = '\t', row.names = F, col.names = F)`

### 5.6 Rcode6-_ScreePlot (Generating Scree Plot, Fig. 6)

*File required:* Output file containing the first 20 eigenvalues and the corresponding principal components, e.g., PCAeigen.csv with the first column "Principal.component" and the second column "Eigenvalue."

Generating the scree plot

- `png(ScreePlot.png, unit = in, width = 5, height = 4, res = 300)`
- `eigenval <- read.csv(PCAeigen.csv, header = TRUE)`
- `plot(eigenval$Principal.component, eigenval $Eigenvalue, type = b)`
- `dev.off()`

### 5.7 Rcode7_manQQ (Generating Manhattan and Quantile-Quantile Plots, Fig. 7)

*Files required:* Output files from our association analysis.

(i) DR_CC.glm.logistic (for logistic regression).

(ii) DR_Linear.glm.linear (for linear regression).

Generating the Manhattan plot

- `install.packages(qqman)`
- `library(qqman)`
- `results <- read.table(DR_CC.glm.logistic, head = TRUE)`
- `colnames(results)[1] = CHR` *# giving appropriate column names, scan the column names of the original file before assigning the column names*
- `colnames(results)[2] = BP`
- `colnames(results)[3] = SNP`
- `colnames(results)[15] = P`
- `colnames(results)[6] = A1`
- `png(Manhattanplot.png,unit-in, width = 7, height = 4, res = 300)`
- `manhattan(results, chr = CHR, bp = BP, p = P, snp = SNP, ylim = c(0, 10), suggestiveline = -log10 (5e-06), annotatePval = 5e-6, col = c(blue, purple, green, red))`
- `dev.off()`

Generating the quantile-quantile plot

- `if(!require(BiocManager, quietly = TRUE))`

  `install.packages(BiocManager)`
  `BiocManager::install(GWASTools)`
- `results <- read.table(DR_CC.glm.logistic.hybrid, head = TRUE)`
- `colnames(results)[1] = CHR` *#giving appropriate column names, scan the column names of the original file before assigning the column names*
- `colnames(results)[2] = BP`
- `colnames(results)[3] = SNP`
- `colnames(results)[15] = P`
- `colnames(results)[6] = A1`
- `png(Quantile quantile plot.png, unit = in, width = 6, height = 5, res = 300)`
- `qqPlot(results_log$P, truncate = FALSE, ylim = NULL, thinThreshold = NULL, ci = TRUE, col = blue, cex = 0.6)`
- `legend(topleft, legend = as.expression(lapply(1: 1, function(x)bquote(lambda == .(1)))), box.lty = 0)`
- `dev.off()`

## References


1. Klein R, Klein BE, Moss SE, Cruickshanks KJ (1998) The Wisconsin epidemiologic study of diabetic retinopathy: XVII: the 14-year incidence and progression of diabetic retinopathy and associated risk factors in type 1 diabetes. Ophthalmology 105(10):1801–1815
2. Lima VC, Cavalieri GC, Lima MC, Nazario NO, Lima GC (2016) Risk factors for diabetic retinopathy: a case–control study. Int J Retina Vitreous 2(1):1–7
3. Cabrera AP, Monickaraj F, Rangasamy S, Hobbs S, McGuire P, Das A (2020) Do genomic factors play a role in diabetic retinopathy? J Clin Med 9(1):216
4. Klein RJ, Zeiss C, Chew EY, Tsai J-Y, Sackler RS, Haynes C, Henning AK, SanGiovanni JP, Mane SM, Mayne ST (2005) Complement factor H polymorphism in age-related macular degeneration. Science 308(5720):385–389
5. Burdon KP, Macgregor S, Hewitt AW, Sharma S, Chidlow G, Mills RA, Danoy P, Casson R, Viswanathan AC, Liu JZ (2011) Genome-wide association study identifies susceptibility loci for open angle glaucoma at TMCO1 and CDKN2B-AS1. Nat Genet 43(6):574–578
6. Grassi MA, Tikhomirov A, Ramalingam S, Below JE, Cox NJ, Nicolae DL (2011) Genome-wide meta-analysis for severe diabetic retinopathy. Hum Mol Genet 20(12): 2472–2481
7. Kuo JZ, Wong TY, Rotter JI (2014) Challenges in elucidating the genetics of diabetic retinopathy. JAMA Ophthalmol 132(1): 96–107
8. Chang CC, Chow CC, Tellier LC, Vattikuti S, Purcell SM, Lee JJ (2015) Second-generation PLINK: rising to the challenge of larger and richer datasets. Gigascience 4(1):s13015–13742
9. Purcell S, Neale B, Todd-Brown K, Thomas L, Ferreira MA, Bender D, Maller J, Sklar P, De

Bakker PI, Daly MJ (2007) PLINK: a tool set for whole-genome association and population-based linkage analyses. Am J Hum Genet 81(3):559–575

10. Das S, Forer L, Schönherr S, Sidore C, Locke AE, Kwong A, Vrieze SI, Chew EY, Levy S, McGue M (2016) Next-generation genotype imputation service and methods. Nat Genet 48(10):1284–1287
11. Huber W, Carey VJ, Gentleman R, Anders S, Carlson M, Carvalho BS, Bravo HC, Davis S, Gatto L, Girke T (2015) Orchestrating high-throughput genomic analysis with Bioconductor. Nat Methods 12(2):115–121
12. Gogarten SM, Bhangale T, Conomos MP, Laurie CA, McHugh CP, Painter I, Zheng X, Crosslin DR, Levine D, Lumley T (2012) GWASTools: an R/Bioconductor package for quality control and analysis of genome-wide association studies. Bioinformatics 28(24): 3329–3331
13. Pruim RJ, Welch RP, Sanna S, Teslovich TM, Chines PS, Gliedt TP, Boehnke M, Abecasis GR, Willer CJ (2010) LocusZoom: regional visualization of genome-wide association scan results. Bioinformatics 26(18):2336–2337
14. Watanabe K, Taskesen E, Van Bochoven A, Posthuma D (2017) Functional mapping and annotation of genetic associations with FUMA. Nat Commun 8(1):1–11
15. Li H (2011) A statistical framework for SNP calling, mutation discovery, association mapping and population genetical parameter estimation from sequencing data. Bioinformatics 27(21):2987–2993
16. Bush WS, Moore JH (2012) Chapter 11: genome-wide association studies. PLoS Comput Biol 8(12):e1002822
17. Anttila V, Kallela M, Oswell G, Kaunisto MA, Nyholt DR, Hämäläinen E, Havanka H, Ilmavirta M, Terwilliger J, Sobel E (2006) Trait components provide tools to dissect the genetic susceptibility of migraine. Am J Hum Genet 79(1):85–99
18. Edwards AO, Ritter R III, Abel KJ, Manning A, Panhuysen C, Farrer LA (2005) Complement factor H polymorphism and age-related macular degeneration. Science 308(5720):421–424
19. Solomon SD, Goldberg MF (2019) ETDRS grading of diabetic retinopathy: still the gold standard? Ophthalmic Res 62(4):185–190
20. Ting DSW, Cheung GCM, Wong TY (2016) Diabetic retinopathy: global prevalence, major risk factors, screening practices and public health challenges: a review. Clin Exp Ophthalmol 44(4):260–277
21. MacRae CA, Vasan RS (2011) Next-generation genome-wide association studies: time to focus on phenotype? Circ Cardiovasc Genet 4:334–336
22. Craddock N, Kendler K, Neale M, Nurnberger J, Purcell S, Rietschel M, Perlis R, Santangelo SL, Schulze T, Smoller JW (2009) Dissecting the phenotype in genome-wide association studies of psychiatric illness. Br J Psychiatry 195(2):97–99
23. Zhang Q, Cai Z, Lhomme M, Sahana G, Lesnik P, Guerin M, Fredholm M, Karlskov-Mortensen P (2020) Inclusion of endophenotypes in a standard GWAS facilitate a detailed mechanistic understanding of genetic elements that control blood lipid levels. Sci Rep 10(1): 1–14
24. Braff DL (2015) The importance of endophenotypes in schizophrenia research. Schizophr Res 163(1–3):1–8
25. Shtir C, Aldahmesh MA, Al-Dahmash S, Abboud E, Alkuraya H, Abouammoh MA, Nowailaty SR, Al-Thubaiti G, Naim EA, ALYounes B, Binhumaid FS, ALOtaibi AB, Altamimi AS, Alamer FH, Hashem M, Abouelhoda M, Monies D, Alkuraya FS (2016) Exome-based case–control association study using extreme phenotype design reveals novel candidates with protective effect in diabetic retinopathy. Hum Genet 135(2): 193–200. https://doi.org/10.1007/s00439-015-1624-8
26. Sobrin L, Green T, Sim X, Jensen RA, Tai ES, Tay WT, Wang JJ, Mitchell P, Sandholm N, Liu Y (2011) Candidate gene association study for diabetic retinopathy in persons with type 2 diabetes: the candidate gene association resource (CARe). Invest Ophthalmol Vis Sci 52(10): 7593–7602
27. Sun JK, Keenan HA, Cavallerano JD, Asztalos BF, Schaefer EJ, Sell DR, Strauch CM, Monnier VM, Doria A, Aiello LP (2011) Protection from retinopathy and other complications in patients with type 1 diabetes of extreme duration: the Joslin 50-year medalist study. Diabetes Care 34(4):968–974
28. Li Y, Levran O, Kim J, Zhang T, Chen X, Suo C (2019) Extreme sampling design in genetic association mapping of quantitative trait loci using balanced and unbalanced case-control samples. Sci Rep 9(1):1–9
29. Pérez-Gracia JL, Gúrpide A, Ruiz-Ilundain MG, Alegría CA, Colomer R, García-Foncillas J, Melero Bermejo I (2010) Selection of extreme phenotypes: the role of clinical observation in translational research. Clin Transl Oncol 12(3):174–180

30. Amanat S, Requena T, Lopez-Escamez JA (2020) A systematic review of extreme phenotype strategies to search for rare variants in genetic studies of complex disorders. Genes (Basel) 11(9):987
31. Cardon LR, Bell JI (2001) Association study designs for complex diseases. Nat Rev Genet 2(2):91–99
32. Liu C, Chen G, Bentley AR, Doumatey A, Zhou J, Adeyemo A, Yang J, Rotimi C (2019) Genome-wide association study for proliferative diabetic retinopathy in Africans. NPJ Genom Med 4(1):1–7
33. Pollack S, Igo RP, Jensen RA, Christiansen M, Li X, Cheng C-Y, Ng MC, Smith AV, Rossin EJ, Segrè AV (2019) Multiethnic genome-wide association study of diabetic retinopathy using liability threshold modeling of duration of diabetes and glycemic control. Diabetes 68(2): 441–456
34. Awata T, Yamashita H, Kurihara S, Morita-Ohkubo T, Miyashita Y, Katayama S, Mori K, Yoneya S, Kohda M, Okazaki Y (2014) A genome-wide association study for diabetic retinopathy in a Japanese population: potential association with a long intergenic non-coding RNA. PLoS One 9(11):e111715
35. Imamura M, Takahashi A, Matsunami M, Horikoshi M, Iwata M, Araki S-i, Toyoda M, Susarla G, Ahn J, Park KH (2021) Genome-wide association studies identify two novel loci conferring susceptibility to diabetic retinopathy in Japanese patients with type 2 diabetes. Hum Mol Genet 30(8):716–726
36. Yang J, Wray NR, Visscher PM (2010) Comparing apples and oranges: equating the power of case-control and quantitative trait association studies. Genet Epidemiol 34(3):254–257
37. Teslovich TM, Musunuru K, Smith AV, Edmondson AC, Stylianou IM, Koseki M, Pirruccello JP, Ripatti S, Chasman DI, Willer CJ (2010) Biological, clinical and population relevance of 95 loci for blood lipids. Nature 466(7307):707–713
38. Klein R, Klein BE, Moss SE, Davis MD, DeMets DL (1984) The Wisconsin epidemiologic study of diabetic retinopathy: II. Prevalence and risk of diabetic retinopathy when age at diagnosis is less than 30 years. Arch Ophthalmol 102(4):520–526
39. Control D, Group CTR (1995) The relationship of glycemic exposure (HbA1c) to the risk of development and progression of retinopathy in the diabetes control and complications trial. Diabetes 44(8):968–983
40. Spencer CC, Su Z, Donnelly P, Marchini J (2009) Designing genome-wide association studies: sample size, power, imputation, and the choice of genotyping chip. PLoS Genet 5(5):e1000477
41. Abhary S, Hewitt AW, Burdon KP, Craig JE (2009) A systematic meta-analysis of genetic association studies for diabetic retinopathy. Diabetes 58(9):2137–2147. https://doi.org/10.2337/db09-0059
42. Purcell S, Cherny SS, Sham PC (2003) Genetic power calculator: design of linkage and association genetic mapping studies of complex traits. Bioinformatics 19(1):149–150
43. Psifidi A, Dovas CI, Bramis G, Lazou T, Russel CL, Arsenos G, Banos G (2015) Comparison of eleven methods for genomic DNA extraction suitable for large-scale whole-genome genotyping and long-term DNA banking using blood samples. PLoS One 10(1): e0115960
44. Consortium WTCC (2007) Genome-wide association study of 14,000 cases of seven common diseases and 3,000 shared controls. Nature 447(7145):661
45. Bhat S, Curach N, Mostyn T, Bains GS, Griffiths KR, Emslie KR (2010) Comparison of methods for accurate quantification of DNA mass concentration with traceability to the international system of units. Anal Chem 82(17):7185–7192
46. Ragoussis J (2009) Genotyping technologies for genetic research. Annu Rev Genomics Hum Genet 10:117–133
47. Guo Y, He J, Zhao S, Wu H, Zhong X, Sheng Q, Samuels DC, Shyr Y, Long J (2014) Illumina human exome genotyping array clustering and quality control. Nat Protoc 9(11): 2643–2662
48. Ledford H (2011) Paper on genetics of longevity retracted. Nature. https://doi.org/10.1038/news.2011.429
49. Lin D-Y, Tao R, Kalsbeek WD, Zeng D, Gonzalez F II, Fernández-Rhodes L, Graff M, Koch GG, North KE, Heiss G (2014) Genetic association analysis under complex survey sampling: the Hispanic Community Health Study/Study of Latinos. Am J Hum Genet 95(6): 675–688
50. Karssen LC, van Duijn CM, Aulchenko YS (2016) The GenABEL Project for statistical genomics. F1000Research 5:914
51. Cardon LR, Palmer LJ (2003) Population stratification and spurious allelic association. Lancet 361(9357):598–604
52. Price AL, Patterson NJ, Plenge RM, Weinblatt ME, Shadick NA, Reich D (2006) Principal components analysis corrects for stratification

in genome-wide association studies. Nat Genet 38(8):904–909

53. Consortium GP, Auton A, Brooks L, Durbin R, Garrison E, Kang H (2015) A global reference for human genetic variation. Nature 526(7571):68–74
54. Rhead B, Karolchik D, Kuhn RM, Hinrichs AS, Zweig AS, Fujita PA, Diekhans M, Smith KE, Rosenbloom KR, Raney BJ (2010) The UCSC genome browser database: update 2010. Nucleic Acids Res 38(suppl_1):D613–D619
55. Marchini J, Howie B (2010) Genotype imputation for genome-wide association studies. Nat Rev Genet 11(7):499–511
56. Nguyen LH, Holmes S (2019) Ten quick tips for effective dimensionality reduction. PLoS Comput Biol 15(6):e1006907
57. Ledesma RD, Valero-Mora P, Macbeth G (2015) The scree test and the number of factors: a dynamic graphics approach. Span J Psychol 18:E11
58. Devlin B, Roeder K (1999) Genomic control for association studies. Biometrics 55(4): 997–1004
59. McCaw ZR, Lane JM, Saxena R, Redline S, Lin X (2020) Operating characteristics of the rank-based inverse normal transformation for quantitative trait analysis in genome-wide association studies. Biometrics 76(4):1262–1272
60. Cuellar-Partida G, Renteria ME, MacGregor S (2015) LocusTrack: integrated visualization of GWAS results and genomic annotation. Source Code Biol Med 10(1):1–4
61. He F, Ding S, Wang H, Qin F (2020) IntAssoPlot: an r package for integrated visualization of genome-wide association study results with gene structure and linkage disequilibrium matrix. Front Genet 11:260
62. Xiao R, Boehnke M (2009) Quantifying and correcting for the winner's curse in genetic association studies. Genet Epidemiol 33(5): 453–462
63. Shang X, Zhu Z, Huang Y, Zhang X, Wang W, Shi D, Jiang Y, Yang X, He M (2021) Associations of ophthalmic and systemic conditions with incident dementia in the UK biobank. Br J Ophthalmol 107(2):275–282
64. Chua SYL, Thomas D, Allen N, Lotery A, Desai P, Patel P, Muthy Z, Sudlow C, Peto T, Khaw PT (2019) Cohort profile: design and methods in the eye and vision consortium of UK biobank. BMJ Open 9(2):e025077
65. Arloth J, Eraslan G, Andlauer TF, Martins J, Iurato S, Kühnel B, Waldenberger M, Frank J, Gold R, Hemmer B (2020) DeepWAS: multivariate genotype-phenotype associations by directly integrating regulatory information using deep learning. PLoS Comput Biol 16(2):e1007616
66. Kwon O-S, Hong M, Kim T-H, Hwang I, Shim J, Choi E-K, Lim HE, Yu HT, Uhm J-S, Joung B (2022) Genome-wide association study-based prediction of atrial fibrillation using artificial intelligence. Open Heart 9(1): e001898
67. Mieth B, Kloft M, Rodríguez JA, Sonnenburg S, Vobruba R, Morcillo-Suárez C, Farré X, Marigorta UM, Fehr E, Dickhaus T (2016) Combining multiple hypothesis testing with machine learning increases the statistical power of genome-wide association studies. Sci Rep 6(1):1–14
68. Leem S, Jeong H-h, Lee J, Wee K, Sohn K-A (2014) Fast detection of high-order epistatic interactions in genome-wide association studies using information theoretic measure. Comput Biol Chem 50:19–28
69. Paré G, Mao S, Deng WQ (2017) A machine-learning heuristic to improve gene score prediction of polygenic traits. Sci Rep 7(1):1–11
70. Nicholls HL, John CR, Watson DS, Munroe PB, Barnes MR, Cabrera CP (2020) Reaching the end-game for GWAS: machine learning approaches for the prioritization of complex disease loci. Front Genet 11:350
71. Grover S, Murthy RK, Brar VS, Chalam KV (2010) Comparison of retinal thickness in normal eyes using stratus and spectralis optical coherence tomography. Invest Ophthalmol Vis Sci 51(5):2644–2647
72. Smith ML, Baggerly KA, Bengtsson H, Ritchie ME, Hansen KD (2013) Illuminaio: an open source IDAT parsing tool for illumina microarrays. F1000Research 2:264
73. Marees AT, de Kluiver H, Stringer S, Vorspan F, Curis E, Marie-Claire C, Derks EM (2018) A tutorial on conducting genome-wide association studies: quality control and statistical analysis. Int J Methods Psychiatr Res 27(2):e1608
74. Anderson CA, Pettersson FH, Clarke GM, Cardon LR, Morris AP, Zondervan KT (2010) Data quality control in genetic case-control association studies. Nat Protoc 5(9): 1564–1573
75. Wittke-Thompson JK, Pluzhnikov A, Cox NJ (2005) Rational inferences about departures from Hardy-Weinberg equilibrium. Am J Hum Genet 76(6):967–986
76. Pe'er I, Yelensky R, Altshuler D, Daly MJ (2008) Estimation of the multiple testing burden for genomewide association studies of nearly all common variants. Genet Epidemiol 32(4):381–385

# Chapter 6

# Single-Cell RNA Sequencing (scRNA-Seq) Data Analysis of Retinal Homeostasis and Degeneration of Microglia

**Madhu Sudhana Saddala, Srilatha Mundla, Naina Patyal, and Srujanika Dash**

## Abstract

Single-cell RNA sequencing (scRNA-seq) experiment reveals previously unseen molecular features. The number of sequencing procedures and computational data analysis approaches has been increasing rapidly in recent years. This chapter provides a general idea of the single-cell data analysis and visualization. An introduction and practical guidance for the 10× sequencing data analysis and visualization are presented. Basic data analysis approaches are highlighted, followed by quality control of data, filtering in cell level and gene level, normalization, dimensional reduction, clustering analysis, and marker identification.

**Key words** Microglia, Retinal degeneration, Diabetic retinopathy, scRNA-Seq, Visualization

## 1 Introduction

Diabetic retinopathy (DR) is the leading cause of vision loss in individuals 20–75 years of age. It remains a frightening prospect to patients and physicians as the causes remain unclear. While DR has been described classically as a microvascular disease, recent evidence suggests that changes to retinal microglia are an early feature of retinopathy [1–3].

The retinal microglia cells showed two different activation forms, pro-inflammatory (M1) and anti-inflammatory (M2) by sensome [4, 5]. The sensome signals are promote the microglial cells to activate numerous pathological signals like secretion of neurotoxins, cytokines, and chemokines, local proliferation, migration, and enhanced phagocytosis. These responses and the loss of restrictive switch mechanisms may contribute to retinal tissue impairment and pro-apoptotic actions in retinal degeneration (RD) [6]. The study of microglia at different stages of diabetes, and its interaction with peripheral leukocytes and retinal cells, is

Guei-Sheung Liu and Jiang-Hui Wang (eds.), *Diabetic Retinopathy: Methods and Protocols*, Methods in Molecular Biology, vol. 2678, https://doi.org/10.1007/978-1-0716-3255-0_6,

crucial to provide insights into mechanisms of damage and repair [7]. The role of microglia in the diabetic retina and normal retina is essential in comprehending the inflammatory components during disease progression.

A standard single-cell sequencing workflow involves initial tissue and cell preparation, cell capturing and library preparation, and sequencing and raw data processing, along with visualization and downstream analyses. Various protocols are available for the preparation of single-cell suspensions because the optimal procedure differs for each tissue and cell type to be isolated [8, 9]. A large number of data analysis platforms (web-based, stand-alone workflows, or integrated into computational frameworks like R or Python) are available to analyze single cell RNA sequencing (scRNA-seq) data. There are commercially available analysis software packages, some of which are developed by single-cell sequencing companies and service providers, such as Cell Ranger and Loupe Cell Browser (10× Genomics) (https://support.10xgenomics.com/genome-exome/software/visualization/latest/what-is-loupe), as well as SeqGeq (BD Bioscience, https://www.bdbiosciences.com/en-us). Others are designed by companies specializing in software solutions, such as Partek Flow (Partek) (https://www.partek.com/partek-flow/). While commercially available packages are user-friendly, open-source analysis packages are usually more powerful, transparent, reproducible, and flexible. Bioconductor, an open-development software project for the analysis of high-throughput genomics data, provides powerful analysis tools, such as Scanpy [10] or Scater [11]. Currently, the R package Seurat [12] is one of the most popular toolboxes for general single-cell sequencing analysis, consistently performing well in benchmarking studies.

In the current study, we use the Emily et al. 2019 datasets for the scRNA-seq data analysis and performed quality control, normalization, data integration, correction, feature selection of data, dimensionality reduction, clustering, marker identification, and visualization.

## 2 Materials and Methods

### *2.1 Microglia scRNA-Seq Data Download*

1. The accession number for the single-cell RNA sequencing data report in this paper is GEO: GSE126783. Here, manually download the single-cell RNA-seq (scRNA-seq) identifies a unique microglia subtype associated with retinal degeneration (https://www.ncbi.nlm.nih.gov/geo/query/acc.cgi?acc=GSE126783) [13].
2. Generally, all single-cell RNA-seq datasets, regardless of technology or pipeline, contain three files: (1) a file with the gene IDs,

representing all genes; (2) a file with the cell IDs, representing all cells; and (3) a matrix of counts per gene for every cell.

3. We can create a count matrix using these files. However, instead of creating a standard count matrix, we create a sparse matrix to improve the amount of space, memory, and CPU required to work with our huge count matrix.
4. We use the readMM function from the Matrix package to turn our standard matrix into a sparse matrix. The genes.tsv file should correspond to the genes or row names of the matrix, while barcodes.tsv corresponds to the cells or columns.
5. We load the necessary libraries.
   #Load libraries:

   library(SingleCellExperiment)
   library(Seurat)
   library(tidyverse)
   library(Matrix)
   library(scales)
   library(cowplot)
   library(RCurl)
6. The For loop uses for iterating over the two sample "files" and execute two commands for each sample: (1) read in the count data (Read10X()) and (2) create the Seurat objects from the read in data (CreateSeuratObject()).

# Create each individual Seurat object for every sample

```
for (file in c("Control", "LD")){
 seurat_data <- Read10X(data.dir = paste0("D:/madhu/huang/
microglia/", file))
seurat_obj <- CreateSeuratObject(counts = seurat_data,
min.features = 100,
project = file)
assign(file, seurat_obj)}
```

7. Next, we merge these objects together into a single Seurat object. This will make it easier to run the QC (quality control) steps for both sample groups together and enable us to easily compare the data quality for all the samples.

# Create a merge Seurat object

```
merged_seurat <- merge(x = Control,
y = c(LD),
add.cell.id = c("Control", "LD"), project = "Microglia")
```

### 2.2 Quality Control

1. We calculate some additional metrics such as the number of genes detected per UMI and mitochondrial ratio, respectively.

# Add number of genes per UMI for each cell to metadata

```
merged_seurat$log10GenesPerUMI < log10(merged_seurat$nFea-
ture_RNA) / log10(merged_seurat$nCount_RNA)
```

# Compute percent mito ratio

```
merged_seurat$mitoRatio <- PercentageFeatureSet(object = mer-
ged_seurat, pattern = "^MT-")
merged_seurat$mitoRatio <- merged_seurat@meta.data$mitoRatio
/ 100
```

2. We create the metadata dataframe by extracting the meta.data slot from the Seurat object.

# Create metadata dataframe

```
metadata <- merged_seurat@meta.data
```

# Add cell IDs to metadata

```
metadata$cells <- rownames(metadata)
```

# Rename columns

```
metadata <- metadata %>%
dplyr::rename(seq_folder = orig.ident,
nUMI = nCount_RNA,
nGene = nFeature_RNA)
```

3. Now, let's get sample names for each of the cells based on the cell prefix.

# Create sample column

```
metadata$sample <- NA
metadata$sample[which(str_detect(metadata$cells, "^Con-
trol_"))] <- "Control"
metadata$sample[which(str_detect(metadata$cells, "^LD_"))]
<- "LD"
```

4. To generate the cell counts, UMI counts per cell, genes detect per cell metrics, mitochondrial counts ratio and UMIs vs. genes detect to assess, we can explore them with visualizations.

# Visualize the number of cell counts per sample

```
metadata %>%
ggplot(aes(x=sample, fill=sample)) +
geom_bar() +
theme_classic() +
 theme(axis.text.x = element_text(angle = 45, vjust =
1, hjust=1)) +
theme(plot.title = element_text(hjust=0.5, face="bold")) +
ggtitle("NCells")
```

# Visualize the number UMIs/transcripts per cell

```
metadata %>%
ggplot(aes(color=sample, x=nUMI, fill= sample)) +
geom_density(alpha = 0.2) +
scale_x_log10() +
theme_classic() +
ylab("Cell density") +
geom_vline(xintercept = 500)
```

# Visualize the distribution of genes detected per cell via histogram

```
metadata %>%
ggplot(aes(color=sample, x=nGene, fill= sample)) +
geom_density(alpha = 0.2) +
theme_classic() +
scale_x_log10() +
geom_vline(xintercept = 300)
```

# Visualize the correlation between genes detected and number of UMIs and determine whether strong presence of cells with low numbers of genes/UMIs

```
metadata %>%
ggplot(aes(x=nUMI, y=nGene, color=mitoRatio)) +
geom_point() +
scale_colour_gradient(low = "gray90", high = "black") +
stat_smooth(method=lm) +
scale_x_log10() +
scale_y_log10() +
theme_classic() +
geom_vline(xintercept = 500) +
geom_hline(yintercept = 250) +
facet_wrap(~sample)
```

# Visualize the distribution of mitochondrial gene expression detected per cell

```
metadata %>%
ggplot(aes(color=sample, x=mitoRatio, fill=sample)) +
geom_density(alpha = 0.2) +
scale_x_log10() +
theme_classic() +
geom_vline(xintercept = 0.2)
```

### *2.3 Filtering*

1. We visualize the abovementioned metrics; we can decide on the thresholds to apply which will result in the removal of low-quality cells. We used the following thresholds (Fig. 1).

   nUMI >500

   nGene >250

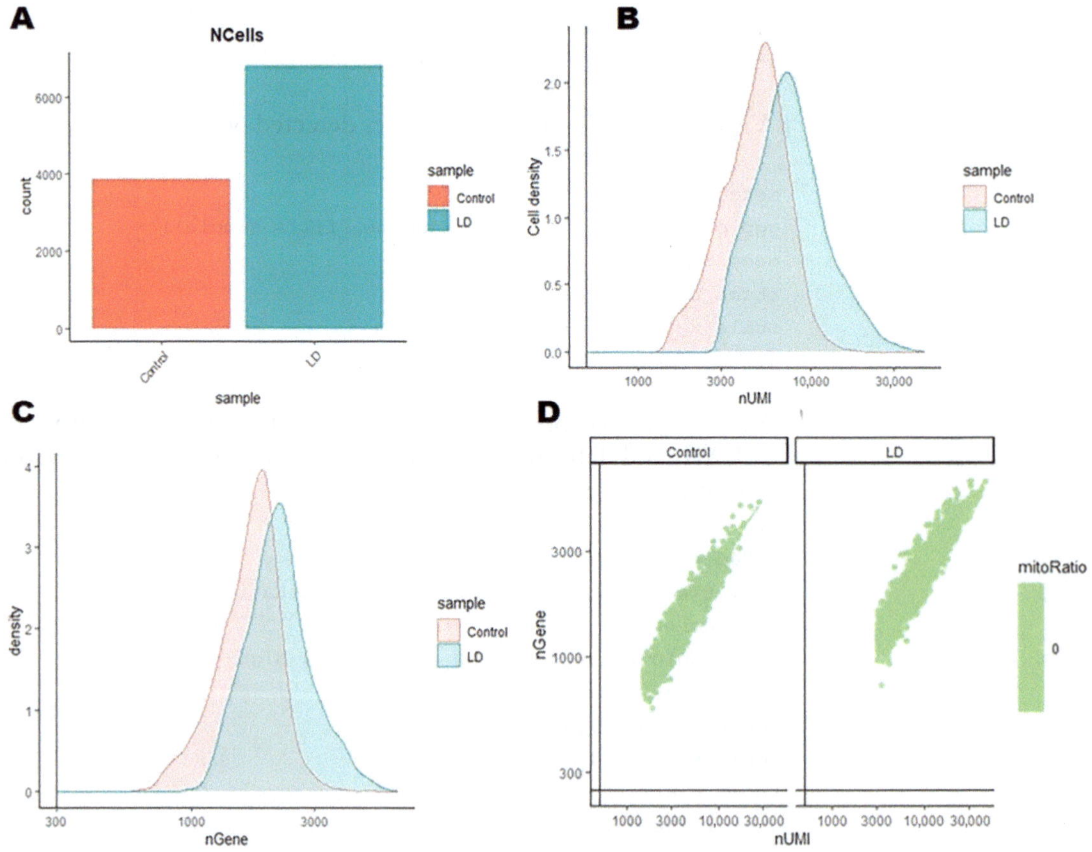


**Fig. 1** Quality control of scRNA-seq data. (**a**) Visualize the number of cell counts per sample. (**b**) Visualize the number UMIs/transcripts per cell. (**c**) Visualize the distribution of genes detected per cell via histogram. (**d**) Visualize of the correlation between genes detected and number of UMIs and determine whether strong presence of cells with low numbers of genes/UMIs

log10GenesPerUMI > 0.8

mitoRatio <0.2

#Cell-level filtering

# Filter out low quality reads using select thresholds—these will change with experiment

```
filtered_seurat <- subset(x = merged_seurat,
subset= (nUMI >= 500) &
(nGene >= 250) &
(log10GenesPerUMI > 0.80) &
(mitoRatio < 0.20))
```

2. Within our data, we have many genes with zero counts. These genes can dramatically reduce the average expression for a cell and so we remove them from our data. First, we remove genes that have zero expression in all cells. For our data, we choose to keep only genes which are expressed in five or more cells.

#Gene-level filtering

# Output a logical vector for every gene on whether the more than zero counts per cell

# Extract counts

```
counts <- GetAssayData(object = filtered_seurat, slot =
"counts")
```

# Output a logical vector for every gene on whether the more than zero counts per cell.

```
nonzero <- counts > 0
```

# Sums all TRUE values and returns TRUE if more than 10 TRUE values per gene

```
keep_genes <- Matrix::rowSums(nonzero) >= 5
```

# Only keeping those genes express in more than 5 cells.

```
filtered_counts <- counts[keep_genes, ]
```

# Reassign to filtered Seurat object

```
filtered_seurat <- CreateSeuratObject(filtered_counts, meta.
data = filtered_seurat@meta.data)
```

### 2.4 Count Normalization and Principal Component Analysis

1. The first step in the analysis is to normalize the raw counts to account for differences in sequencing depth per cell for each sample. Seurat recently introduces a new method for the normalization and variance stabilization of scRNA-seq data called sctransform.

# Normalize the counts

```
seurat_phase <- NormalizeData(filtered_seurat)
```

2. Once the data is normalized for sequencing depth, we can assign each cell a score, based on its expression of G2/M and S phase markers. We provide a list of human cell cycle markers for you to download. Right-click on this link (https://www.dropbox.com/s/hus4mrkueh1tfpr/cycle.rda?dl=1) to "Save link as . . ." directly into your data directory.

# Load cell cycle markers

```
load("data/cycle.rda")
```

# Score cells for cell cycle

```
seurat_phase <- CellCycleScoring(seurat_phase,
g2m.features = g2m_genes,
s.features = s_genes)
```

3. After scoring the cells for cell cycle, we would like to determine whether cell cycle is a major source of variation in our dataset using PCA. To perform PCA, we need to first choose the most variable features and then scale the data. Since highly expressed genes exhibit the highest amount of variation and we don't want our "highly variable genes" only to reflect high expression, we need to scale the data to scale variation with expression level (Fig. 2) (*see* **Note 1**).

# Identify the most variable genes

```
seurat_phase <- FindVariableFeatures(seurat_phase,
selection.method = "vst",
nfeatures = 2000,
verbose = FALSE)
```

# Scale the counts

```
seurat_phase <- ScaleData(seurat_phase)
```

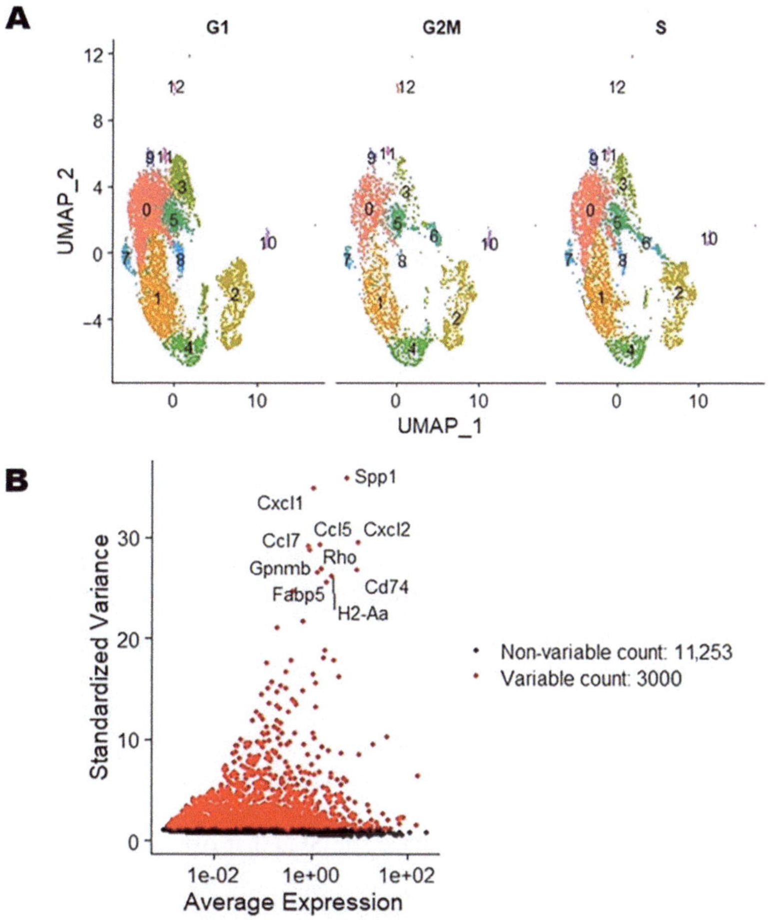


**Fig. 2** Single-cell RNA-seq clustering analysis, normalization, and variance stabilization. (**a**) Cell cycle scores and phases assigned to cells. (**b**) Visualization of variable features: red color indicates highly variable genes, and black color indicates non-variable genes

4. Now, we can perform the principal component analysis (PCA) and plot the top PCs.

# Perform PCA

```
seurat_phase <- RunPCA(seurat_phase)
```

# Plot the PCA color by cell cycle phase

```
DimPlot(seurat_phase,
reduction = "pca",
group.by= "Phase",
split.by = "Phase")
```

5. Now, to perform the cell cycle scoring by sctransform method on all samples. This may take some time (~10 min).

# Split Seurat object by condition to perform cell cycle scoring and SCT on all samples

```
split_seurat <- SplitObject(filtered_seurat, split.by = "sam-
ple")
split_seurat <- split_seurat[c("Control", "LD")]
for (i in 1:length(split_seurat)) {
split_seurat[[i]] <- NormalizeData(split_seurat[[i]], verbose
= TRUE)
split_seurat[[i]] <- CellCycleScoring(split_seurat[[i]], g2m.
features=g2m_genes, s.features=s_genes)
split_seurat[[i]] <- SCTransform(split_seurat[[i]], vars.to.
regress = c("mitoRatio"))}
```

6. We specify that we use all of the 3000 most variable genes identified by SCTransform for the integration.

# Select the most variable features to use for integration

```
integ_features <- SelectIntegrationFeatures(object.list =
split_seurat,
nfeatures = 3000)
```

# Prepare the SCT list object for integration

```
split_seurat <- PrepSCTIntegration(object.list = split_seur-
at,
anchor.features = integ_features)
```

7. We perform canonical correlation analysis (CCA), find the best buddies or anchors, and filter incorrect anchors.

# Find best buddies—can take a while to run

```
integ_anchors <- FindIntegrationAnchors(object.list = split_-
seurat,
normalization.method = "SCT",
anchor.features = integ_features)
```

# Integrate across conditions

```
seurat_integrated <- IntegrateData(anchorset = integ_anchors,
normalization.method = "SCT")
```

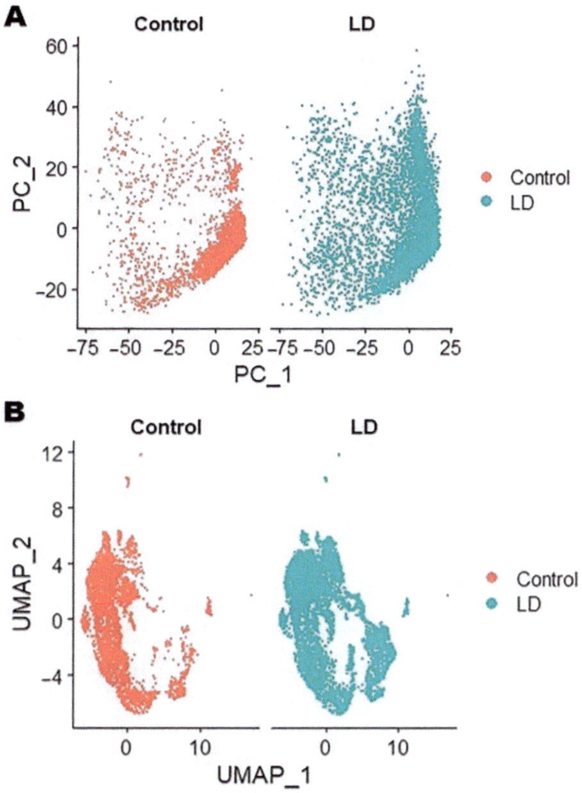


**Fig. 3** Single-cell RNA-seq visualization with (**a**) PCA and (**b**) UMAPs

### 2.5 Visualization

After integration, to visualize the integrated data, we use dimensionality reduction techniques, such as PCA and Uniform Manifold Approximation and Projection (UMAP) (Fig. 3).

# Run PCA

```
seurat_integrated <- RunPCA(object = seurat_integrated)
```

# Plot PCA

```
PCAPlot(seurat_integrated,
 split.by = "sample")
```

# Run UMAP

```
seurat_integrated <- RunUMAP(seurat_integrated,
 dims = 1:40,
 reduction = "pca")
```

# Plot UMAP.

```
DimPlot(seurat_integrated)
```

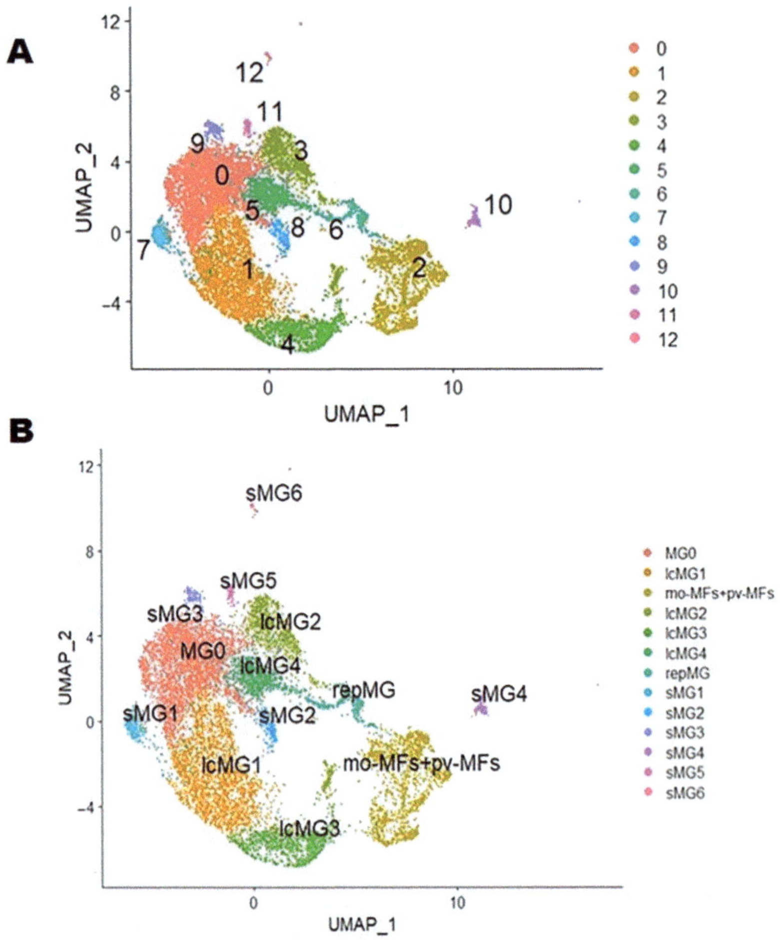


**Fig. 4** Single-cell RNA-seq clustering and assigned cell type clusters. (**a**) Assign numbers to identified clusters and (**b**) assign identity of clusters labels, exploring known microglia cell type markers

### 2.6 Clustering Analysis

1. Seurat uses a graph-based clustering approach, which embeds cells in a graph structure, using a K-nearest neighbor (KNN) graph (by default), with edges drawn between cells with similar gene expression patterns. Then, it attempts to partition this graph into highly interconnected quasi-cliques or communities (Fig. 4).

# Determine the K-nearest neighbor graph

```
seurat_integrated <- FindNeighbors(object = seurat_inte-
grated,
 dims = 1:30)
```

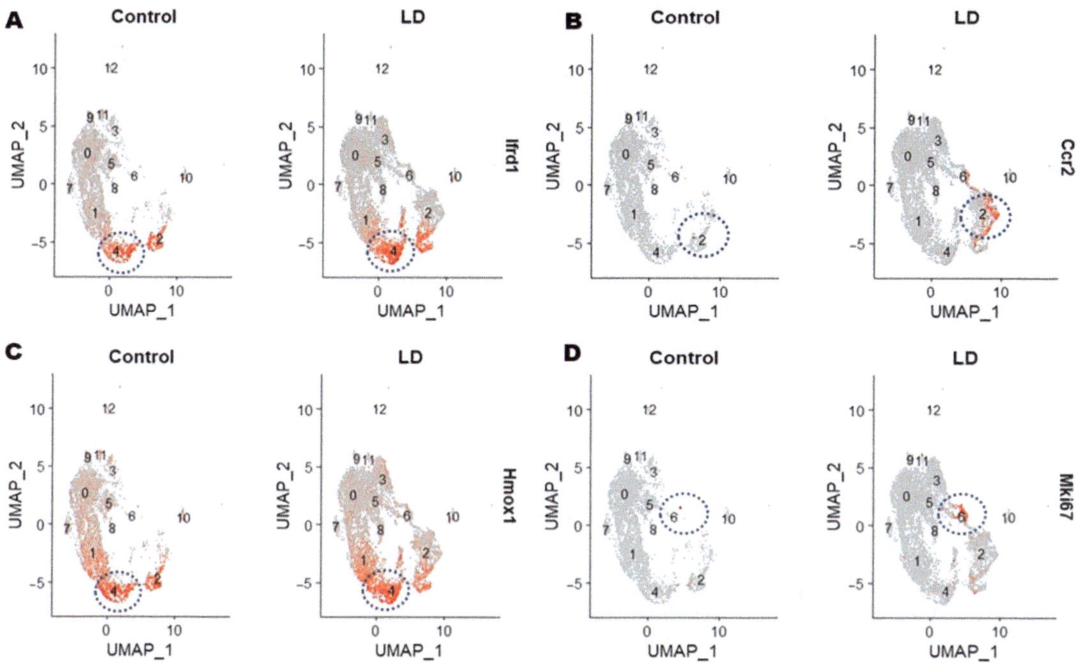


**Fig. 5** Exploring known microglia cell type markers (cluster biomarkers). (**a**) Ifrd1 marker gene, (**b**) Ccr2 marker gene, (**c**) Hmox1 marker gene, and (**d**) Mki67 marker gene

# Determine the clusters for various resolutions

```
seurat_integrated <- FindClusters(object = seurat_integrated,
resolution = c(0.2))
head(seurat_integrated)
 Idents(object = seurat_integrated) <- "integra-
ted_snn_res.0.2"
```

# Plot the UMAP.

```
DimPlot(seurat_integrated,
reduction = "umap",
label = TRUE,
label.size = 6)
```

### 2.7 Cell Type Markers

1. Here, we check the microglia known markers for cell types (Fig. 5).

```
FeaturePlot(seurat_integrated,
reduction = "umap",
features = c("Ifrd1"),
sort.cell = TRUE,
min.cutoff = 'q10',
```

```
label = TRUE)
FeaturePlot(seurat_integrated,
reduction = "umap",
features = c("Sparc"),
sort.cell = TRUE,
min.cutoff = 'q10',
label = TRUE)

FeaturePlot(seurat_integrated,
reduction = "umap",
features = c("Lag3"),
sort.cell = TRUE,
min.cutoff = 'q10',
label = TRUE)
FeaturePlot(seurat_integrated,
reduction = "umap",
features = c("Ccr2"),
sort.cell = TRUE,
min.cutoff = 'q10',
label = TRUE)
```

2. We surmise the cell types of the different clusters and plot the cells with cell type labels.

```
seurat_integrated <- RenameIdents(object = seurat_integrated,
"0" = "MG0",
"1" = "lcMG1",
"2" = "mo-MFs+pv-MFs",
"3" = "lcMG2",
"4" = "lcMG3",
"5" = "lcMG4",
"6" = "repMG",
"7" = "sMG1",
"8" = "sMG2",
"9" = "sMG3",
"10" = "sMG4",
"11" = "sMG5",
"12" = "sMG6")
# Plot the UMAP
DimPlot(object = seurat_integrated,
reduction = "umap",
label = TRUE,
label.size = 6,
repel = TRUE)
# Save final R object
write_rds(seurat_integrated,
path ="D:/madhu/microglia/seurat_labelled.rds")
```

## 3 Notes

1. For the selection.method and nfeatures arguments, the values specified are the default settings. Therefore, you do not necessarily need to include these in your code. We include it here for transparency and inform you what you are using.

## Acknowledgments

The authors would like to thank Chen Xiaoni and Juthika (University of Missouri, Columbia, Missouri, USA) for her language assistance.

### References

1. Lieth E, Gardner TW, Barber AJ, Antonetti DA, Penn State Retina Research G (2000) Retinal neurodegeneration: early pathology in diabetes. Clin Exp Ophthalmol 28:3–8. https://doi.org/10.1046/j.1442-9071.2000.00222.x
2. Zeng XX, Ng YK, Ling EA (2000) Neuronal and microglial response in the retina of streptozotocin-induced diabetic rats. Vis Neurosci 17:463–471. https://doi.org/10.1017/s0952523800173122
3. Saddala MS, Yang X, Tang S, Huang H (2021) Transcriptome-wide analysis reveals core sets of transcriptional regulators of sensome and inflammation genes in retinal microglia. Genomics 113:3058–3071. https://doi.org/10.1016/j.ygeno.2021.07.001
4. Cho BP, Song DY, Sugama S, Shin DH, Shimizu Y, Kim SS, Kim YS, Joh TH (2006) Pathological dynamics of activated microglia following medial forebrain bundle transection. Glia 53:92–102. https://doi.org/10.1002/glia.20265
5. Barcia C, Ros CM, Annese V, Gomez A, Ros-Bernal F, Aguado-Llera D, Martinez-Pagan ME, de Pablos V, Fernandez-Villalba E, Herrero MT (2012) IFN-gamma signaling, with the synergistic contribution of TNF-alpha, mediates cell specific microglial and astroglial activation in experimental models of Parkinson's disease. Cell Death Dis 3: e379. https://doi.org/10.1038/cddis.2012.123
6. Ardeljan D, Chan CC (2013) Aging is not a disease: distinguishing age-related macular degeneration from aging. Prog Retin Eye Res 37:68–89. https://doi.org/10.1016/j.preteyeres.2013.07.003
7. Grigsby JG, Cardona SM, Pouw CE, Muniz A, Mendiola AS, Tsin AT, Allen DM, Cardona AE (2014) The role of microglia in diabetic retinopathy. J Ophthalmol 2014:705783. https://doi.org/10.1155/2014/705783
8. Liao J, Yu Z, Chen Y, Bao M, Zou C, Zhang H, Liu D, Li T, Zhang Q, Li J, Cheng J, Mo Z (2020) Single-cell RNA sequencing of human kidney. Sci Data 7:4. https://doi.org/10.1038/s41597-019-0351-8
9. Bykov Y, Kim SH, Zamarin D (2020) Preparation of single cells from tumors for single-cell RNA sequencing. Methods Enzymol 632: 295–308. https://doi.org/10.1016/bs.mie.2019.05.057
10. Wolf FA, Angerer P, Theis FJ (2018) SCANPY: large-scale single-cell gene expression data analysis. Genome Biol 19:15. https://doi.org/10.1186/s13059-017-1382-0
11. McCarthy DJ, Campbell KR, Lun AT, Wills QF (2017) Scater: pre-processing, quality control, normalization and visualization of single-cell RNA-seq data in R. Bioinformatics 33: 1179–1186. https://doi.org/10.1093/bioinformatics/btw777
12. Butler A, Hoffman P, Smibert P, Papalexi E, Satija R (2018) Integrating single-cell transcriptomic data across different conditions,

technologies, and species. Nat Biotechnol 36: 411–420. https://doi.org/10.1038/nbt.4096

13. O'Koren EG, Yu C, Klingeborn M, Wong AYW, Prigge CL, Mathew R, Kalnitsky J, Msallam RA, Silvin A, Kay JN, Bowes Rickman C, Arshavsky VY, Ginhoux F, Merad M, Saban DR (2019) Microglial function is distinct in different anatomical locations during retinal homeostasis and degeneration. Immunity 50: 723–737 e7. https://doi.org/10.1016/j.immuni.2019.02.007

# Chapter 7

# Bioinformatics Tools for Bulk Gene Expression Deconvolution in Diabetic Retinopathy

**Ru Qi Teh, Guei-Sheung Liu, and Jiang-Hui Wang**

## Abstract

Retinal neovascularization is one of the leading causes of vision loss and a hallmark of proliferative diabetic retinopathy (PDR). The immune system is observed to be involved in the pathogenesis of diabetic retinopathy (DR). The specific immune cell type that contributes to retinal neovascularization can be identified via a bioinformatics analysis of RNA sequencing (RNA-seq) data, known as deconvolution analysis. Previous study has identified the infiltration of macrophages in the retina of rats with hypoxia-induced retinal neovascularization and patients with PDR through a deconvolution algorithm, known as CIBERSORTx. Here, we describe the protocols of using CIBERSORTx to perform the deconvolution analysis and downstream analysis of RNA-seq data.

**Key words** Retinal neovascularization, Deconvolution analysis, Bioinformatics, Gene Expression Omnibus, ArrayExpress, CIBERSORTx

## 1 Introduction

Retinal neovascularization is a severe complication of proliferative diabetic retinopathy (PDR) that can lead to visual impairment or blindness [1, 2]. PDR is a classic microvascular complication of diabetes and traditionally characterized as vascular abnormalities [3]. Recent studies have discovered the involvement of the immune system in the pathogenesis of PDR, including complement system and immune cells [3–5]. Moreover, increased cytokines and chemokine were observed in the ocular tissues, such as vitreous and aqueous humor, of both patients and animal models with diabetes [6]. However, the mechanisms by which the immune response is involved in PDR remain unclear, whether it is a local activation in the retina or a systematic complication of diabetes.

There are different types of immune cells, not all cell types contributing to the disease pathogenesis [3]. Previous studies have identified specific immune cells that are relevant to

Guei-Sheung Liu and Jiang-Hui Wang (eds.), *Diabetic Retinopathy: Methods and Protocols*, Methods in Molecular Biology, vol. 2678, https://doi.org/10.1007/978-1-0716-3255-0_7,

inflammation in retinal neovascularization, including mononuclear phagocytes and macrophages [7–10]. Mononuclear phagocyte activity in diabetic retinopathy (DR) leads to retinal degeneration, and the elevation of cytokines, such as Monocyte chemotactic protein 1 (MCP-1)/C-C motif chemokine 2 (CCL2) and vascular endothelial growth factor (VEGF), is associated with neovascularization and the progression of DR [10]. It is difficult to differentiate the infiltrating mononuclear phagocytes and elevated cytokines from regular immune activity. Hence, single-cell RNA sequencing (scRNA-seq) helps to characterize the molecular information of individual cell types and improve the understanding of the pathogenesis of retinal neovascularization, especially in PDR [11].

The study of cell proportions in the bulk tissue is important to understand the quantitative cellular change in the pathogenesis of diseases. Unlike the traditional assessment, such as flow cytometry and immunohistochemistry, that requires complex cell preparation and is labor- and cost-intensive, bulk RNA-seq provides transcriptional information of bulk tissues which include diverse types of RNA species and genomic alterations [12, 13]. The conventional analysis of bulk tissues has been focused on the identification of differentially expressed genes between the healthy and diseased bulk tissues [14]. However, a more popular methodology—deconvolution analysis of transcriptomic data—has been described by recent studies [15]. Deconvolution methodology develops an algorithm to infer cellular proportions with transcriptional data according to a signature matrix featuring quantitative information of cell markers of interest cells [16].

Many computational methods for deconvolution analysis have been previously described, including csSAM, DSA, MMAD, ISOpure, TIMER, BSEQ-sc, CIBERSORT, DeMixT, MuSiC, CPM, and CIBERSORTx [17–30]. Of which, CIBERSORTx provides better performance and versatility in digital cytometry. CIBERSORTx is the "in silico flow cytometry" that reduces the effort of physical cell sorting. CIBERSORTx was developed in Alizadeh and Newman lab to impute cell proportion and gene expression profile from transcriptomic data [30, 31]. The framework utilizes a curated signature matrix from reference cell types to process gene expression data to infer cell types and gene expression at single cell level in group mode and high resolution, respectively [30]. Previous studies by our group have identified the infiltration of M2 macrophages and its significant increase in the retina of a rat model and patients with retinal neovascularization through CIBERSORTx [32].

Here, we will present a protocol of bulk gene expression deconvolution analysis using CIBERSORTx to identify the changes in estimated proportions of immune cells in the retina from a rat model that features retinal neovascularization. The pipeline for deconvolution analysis is shown in Fig. 1.

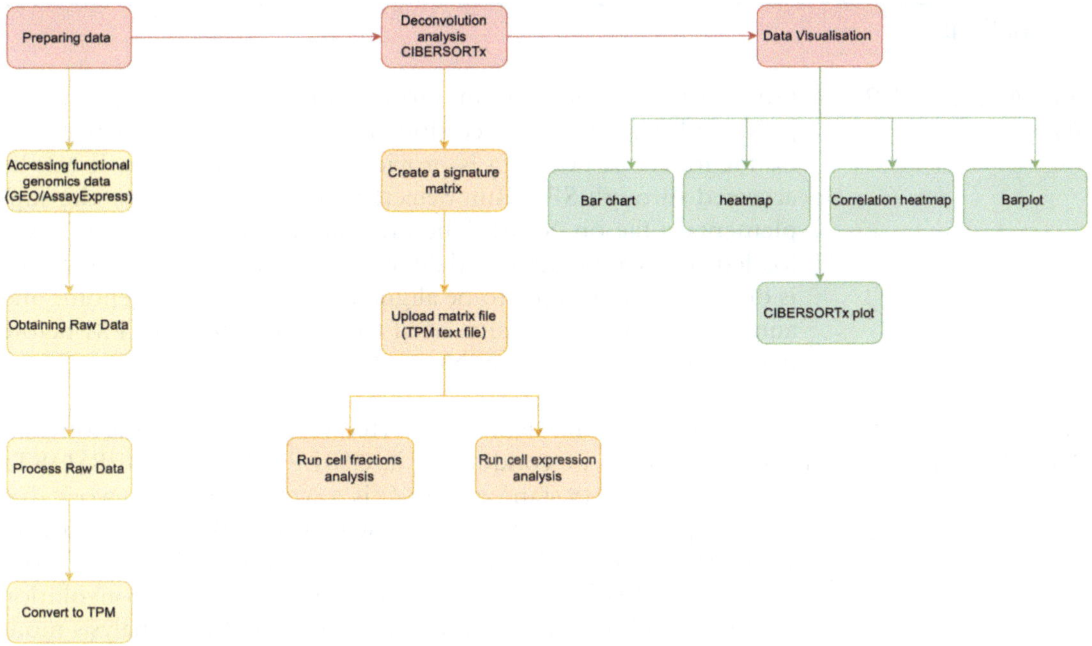


**Fig. 1** Bioinformatics pipeline for deconvolution analysis

## 2 Materials

### *2.1 Online Databases for Functional Genomics Data*

Before analyzing with CIBERSORTx, bulk RNA-seq data in transcripts per million (TPM) value is required. To facilitate the functional genomics society in archiving functional genomics data, two publicly available databases are available to help maintain and facilitate high-throughput data. These databases allow users to share and reuse data for data mining and discovery.

### *2.2 ArrayExpress*

ArrayExpress is an archive for genomic data, developed by the European Molecular Biology Laboratory's European Bioinformatics Institute (EMBL-EBI). ArrayExpress obtains data directly from authors' submissions and imports data from NCBI GEO. The supported data type includes processed data, raw data, and atlas. Both online databases provide RNA-seq data that can be used for deconvolution analysis. To submit the data to CIBERSORTx, the read count text file needs to be formatted in TPM.

### *2.3 Gene Expression Omnibus (GEO)*

GEO (https://www.ncbi.nlm.nih.gov/geo/) is a publicly available repository maintained by the National Center for Biotechnology Information (NCBI) for genomic data from high-throughput genomic technologies, such as microarray and next-generation sequencing [33]. GEO is an archive for raw or processed data and metadata of functional genomic data. In this protocol, we acquired raw data from GEO.

## 3 Methods

### 3.1 Accessing GEO Data

GEO data is submitted in various formats, including raw and processed data. Firstly, accession of the data of interest needs to be identified. If raw data is provided by the submitter, it can be accessed through SRA Run Selector with the link below the supplementary file on NCBI. The raw data of interest can be downloaded via SRA toolkit, cloud delivery, or Galaxy. Once the raw data is downloaded, it needs to be aligned to the reference genome and annotated. The read counts need to be converted to TPM before deconvolution analysis (*see* **Notes 1** and **2**).

### 3.2 Deconvolution Tool: CIBERSORTx

Once the read count data is converted to TPM format, deconvolution analysis can be conducted via CIBERSORTx. CIBERSORTx (https://cibersortx.stanford.edu/) is a web-based framework that is developed in Alizadeh lab and Newman lab [30, 31] (*see* **Note 3**). In order to describe the details of using CIBERSORTx, we take our latest study [34] as an example to showcase the deconvolution analysis with bulk RNA-seq dataset. There are three analysis modules in CIBERSORTx as follows:

1. Create Signature Matrix
2. Impute Cell Fractions
3. Impute Cell Expression

### 3.3 Creating Signature Matrix

To conduct deconvolution analysis of bulk RNA-seq data, a custom signature matrix is required to define the expression profile of the cell type of interest (*see* **Note 4**). The signature matrix serves as the reference profile for downstream analysis. A custom signature matrix can be created from single-cell RNA sequencing analysis or bulk RNA expression profiles. A detailed protocol of establishing a signature matrix can be referred to Binbin Chen's study [35]. In this protocol, we use the predesigned immune cell signature matrix, LM22, a signature matrix file consisting of 547 genes that accurately distinguish 22 mature human hematopoietic populations isolated from peripheral blood or in vitro culture conditions, including 7 T-cell types, naive and memory B cells, plasma cells, NK cells, and myeloid subsets. LM22 was designed and extensively validated on gene expression microarray data but is also applicable to RNA-seq data from both humans and animals [30, 35].

### 3.4 Imputing Cell Fractions

In this module, the proportions of distinct cell subpopulations are enumerated from bulk expression profiles. It deconvolutes bulk RNA-seq data with a signature matrix generated by single cell transcriptomes and sorted cell populations.

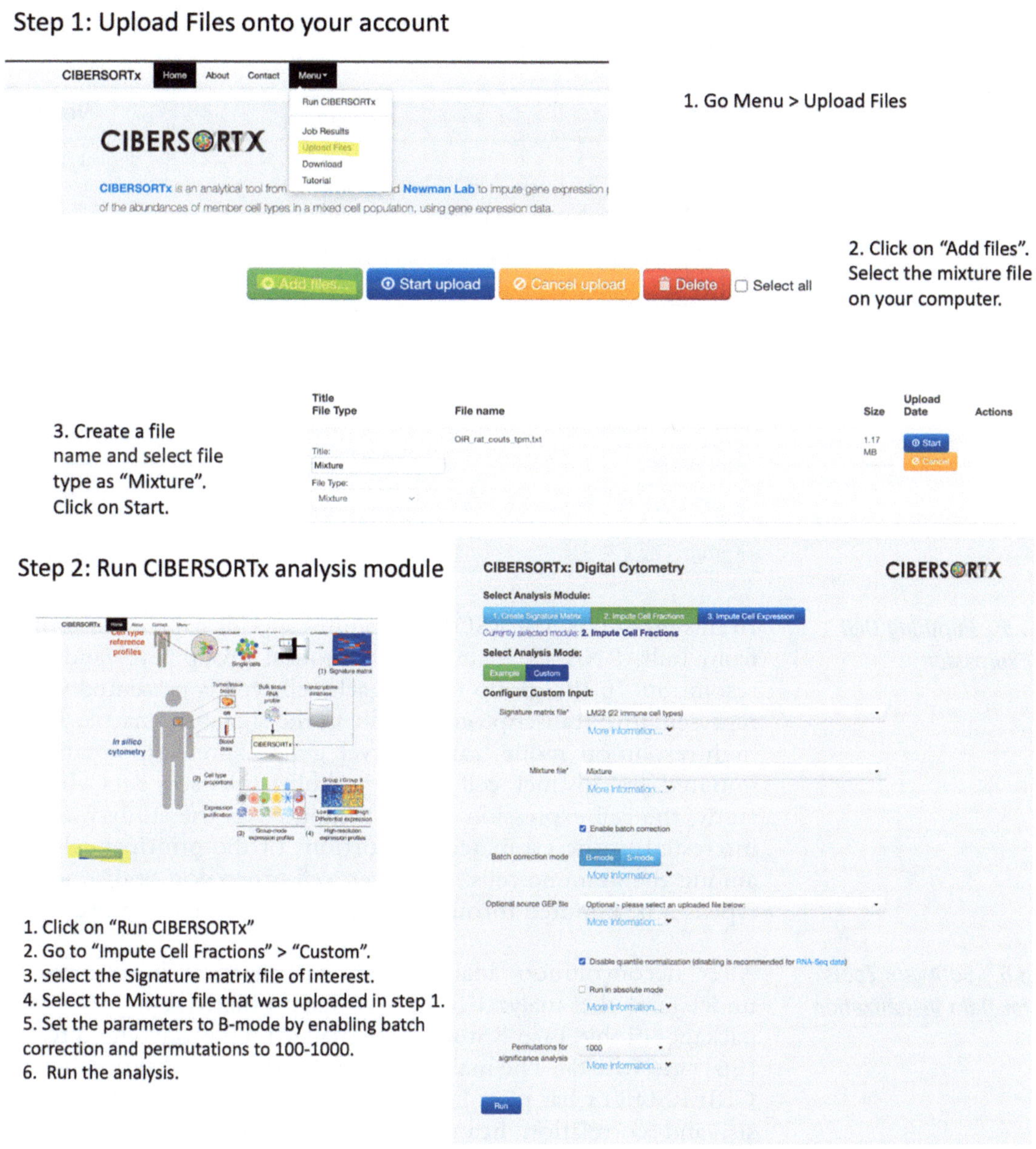


**Fig. 2** Steps for running CIBERSORTx

Firstly, upload the reference mixture file (hereafter referred to as "Mixture.txt") onto CIBERSORTx under "Menu > Upload Files" as shown in Fig. 2, and select "Mixture" as file type (*see* **Note 5**). Secondly, go to the "Run" tab and select "2. Impute Cell Fractions" and select "custom" (Fig. 2). Under the signature matrix file, select "LM22" and the OIR mixture file as the mixture file. For the parameters, enable batch correction mode in B-mode (bulk mode), disable quantile normalization, and set the permutations for significance analysis to 100–1000 (*see* **Note 6**).

**Table 1**
**Versions of R packages**

| Package | Version |
|---|---|
| RColorBrewer | 1.1-2 |
| ggplot2 | 3.3.5 |
| limma | 3.46.0 |
| tidyverse | 1.3.1 |
| pheatmap | 1.0.12 |
| viridis | 0.6.2 |
| corrplot | 0.92 |
| ggthemes | 4.2.4 |
| ggstatsplot | 0.9.1 |
| ggpubr | 0.4.0 |

### 3.5 Imputing Cell Expression

In this module, CIBERSORTx imputes specific expression profiles from bulk RNA-seq data in two modes: group level and high resolution. In the group mode, each cell type is presented with a representative transcriptome profile in the signature matrix. In the high-resolution mode, sample-level gene expression variation is imputed for distinct cell types in bulk RNA-seq data. In this study, the cell expression was not imputed as the study was only interested in the estimated proportions of the principal cell type among the immune cells. However, cell expression of distinct cell type can be imputed through the analysis module (*see* **Note 7**).

### 3.6 Software Tools for Data Visualization

After deconvolution analysis, data visualization is required to understand the analyzed data. We have utilized a number of R packages (Table 1) in RStudio to visualize the deconvoluted results [36] (*see* **Note 8**). The main diagrams for data visualization include CIBERSORTx bar plot, heatmap, bar charts with statistical analysis, and correlation heatmap as shown in our previous study [32]. The code for data visualization is deposited to GitHub (https://github.com/ausloan/CIBERSORTx-OIR).

## 4 Notes

1. The gene symbols for annotations must be consistent throughout the analysis.
2. The text file needs to be in TPM format.
3. An account needs to be created on CIBERSORTx to access the analytical tool. It may take a few business days for the team to approve the account.

4. Custom signature matrices can be created according to the cell types of interest of specific study. Detailed steps are described in the tutorial on the website. Refer to the tutorial (https://cibersortx.stanford.edu/tutorial.php) for further detail.
5. The file name cannot consist of any spaces and symbols, except for "_". Otherwise, the uploaded file cannot be found online. Other features of the analysis module can be found in the tutorial (https://cibersortx.stanford.edu/tutorial.php).
6. If a customized signature matrix is used, enable batch correction to S-mode, and upload the reference file of scRNA-seq in TPM format.
7. Detailed tutorial of the analysis module is described in the tutorial (https://cibersortx.stanford.edu/tutorial.php).
8. The package's name is case-sensitive.

## Acknowledgments

This work was supported by grants from the National Health and Medical Research Council of Australia (GNT1185600). The Centre for Eye Research Australia receives Operational Infrastructure Support from the Victorian Government.

### References

1. Campochiaro PA (2013) Ocular neovascularization. J Mol Med (Berl) 91(3):311–321. https://doi.org/10.1007/s00109-013-0993-5
2. Ishibazawa A, Nagaoka T, Yokota H, Takahashi A, Omae T, Song Y-S, Takahashi T, Yoshida A (2016) Characteristics of retinal neovascularization in proliferative diabetic retinopathy imaged by optical coherence tomography angiography. Invest Ophthalmol Vis Sci 57(14):6247–6255. https://doi.org/10.1167/iovs.16-20210
3. Pan WW, Lin F, Fort PE (2021) The innate immune system in diabetic retinopathy. Prog Retin Eye Res 84:100940
4. Kong D, Gong L, Arnold E, Shanmugam S, Fort PE, Gardner TW, Abcouwer SF (2016) Insulin-like growth factor 1 rescues R28 retinal neurons from apoptotic death through ERK-mediated BimEL phosphorylation independent of Akt. Exp Eye Res 151:82–95
5. Mukai R, Okunuki Y, Husain D, Kim CB, Lambris JD, Connor KM (2018) The complement system is critical in maintaining retinal integrity during aging. Front Aging Neurosci 10:15
6. Rübsam A, Parikh S, Fort PE (2018) Role of inflammation in diabetic retinopathy. Int J Mol Sci 19(4):942
7. Demircan N, Safran B, Soylu M, Ozcan A, Sizmaz S (2006) Determination of vitreous interleukin-1 (IL-1) and tumour necrosis factor (TNF) levels in proliferative diabetic retinopathy. Eye 20(12):1366–1369
8. Murugeswari P, Shukla D, Rajendran A, Kim R, Namperumalsamy P, Muthukkaruppan V (2008) Proinflammatory cytokines and angiogenic and anti-angiogenic factors in vitreous of patients with proliferative diabetic retinopathy and Eales' disease. Retina 28(6):817–824
9. Boss JD, Singh PK, Pandya HK, Tosi J, Kim C, Tewari A, Juzych MS, Abrams GW, Kumar A (2017) Assessment of neurotrophins and inflammatory mediators in vitreous of patients with diabetic retinopathy. Invest Ophthalmol Vis Sci 58(12):5594–5603
10. Akhtar-Schäfer I, Wang L, Krohne TU, Xu H, Langmann T (2018) Modulation of three key innate immune pathways for the most common retinal degenerative diseases. EMBO Mol Med 10(10):e8259

11. Yuan G-C, Cai L, Elowitz M, Enver T, Fan G, Guo G, Irizarry R, Kharchenko P, Kim J, Orkin S, Quackenbush J, Saadatpour A, Schroeder T, Shivdasani R, Tirosh I (2017) Challenges and emerging directions in single-cell analysis. Genome Biol 18(1):84. https://doi.org/10.1186/s13059-017-1218-y
12. Wang Z, Gerstein M, Snyder M (2009) RNA-Seq: a revolutionary tool for transcriptomics. Nat Rev Genet 10(1):57–63. https://doi.org/10.1038/nrg2484
13. Li X, Wang C-Y (2021) From bulk, single-cell to spatial RNA sequencing. Int J Oral Sci 13(1):36. https://doi.org/10.1038/s41368-021-00146-0
14. Rapaport F, Khanin R, Liang Y, Pirun M, Krek A, Zumbo P, Mason CE, Socci ND, Betel D (2013) Comprehensive evaluation of differential gene expression analysis methods for RNA-seq data. Genome Biol 14(9):3158. https://doi.org/10.1186/gb-2013-14-9-r95
15. Jin H, Liu Z (2021) A benchmark for RNA-seq deconvolution analysis under dynamic testing environments. Genome Biol 22(1):102. https://doi.org/10.1186/s13059-021-02290-6
16. Avila Cobos F, Vandesompele J, Mestdagh P, De Preter K (2018) Computational deconvolution of transcriptomics data from mixed cell populations. Bioinformatics 34(11):1969–1979. https://doi.org/10.1093/bioinformatics/bty019
17. Abbas AR, Wolslegel K, Seshasayee D, Modrusan Z, Clark HF (2009) Deconvolution of blood microarray data identifies cellular activation patterns in systemic lupus erythematosus. PLoS One 4(7):e6098. https://doi.org/10.1371/journal.pone.0006098
18. Shen-Orr SS, Tibshirani R, Khatri P, Bodian DL, Staedtler F, Perry NM, Hastie T, Sarwal MM, Davis MM, Butte AJ (2010) Cell type-specific gene expression differences in complex tissues. Nat Methods 7(4):287–289. https://doi.org/10.1038/nmeth.1439
19. Gong T, Hartmann N, Kohane IS, Brinkmann V, Staedtler F, Letzkus M, Bongiovanni S, Szustakowski JD (2011) Optimal deconvolution of transcriptional profiling data using quadratic programming with application to complex clinical blood samples. PLoS One 6(11):e27156. https://doi.org/10.1371/journal.pone.0027156
20. Kuhn A, Thu D, Waldvogel HJ, Faull RL, Luthi-Carter R (2011) Population-specific expression analysis (PSEA) reveals molecular changes in diseased brain. Nat Methods 8(11):945–947. https://doi.org/10.1038/nmeth.1710
21. Qiao W, Quon G, Csaszar E, Yu M, Morris Q, Zandstra PW (2012) PERT: a method for expression deconvolution of human blood samples from varied microenvironmental and developmental conditions. PLoS Comput Biol 8(12):e1002838. https://doi.org/10.1371/journal.pcbi.1002838
22. Ahn J, Yuan Y, Parmigiani G, Suraokar MB, Diao L, Wistuba II, Wang W (2013) DeMix: deconvolution for mixed cancer transcriptomes using raw measured data. Bioinformatics 29(15):1865–1871. https://doi.org/10.1093/bioinformatics/btt301
23. Quon G, Haider S, Deshwar AG, Cui A, Boutros PC, Morris Q (2013) Computational purification of individual tumor gene expression profiles leads to significant improvements in prognostic prediction. Genome Med 5(3):29. https://doi.org/10.1186/gm433
24. Zhong Y, Wan Y-W, Pang K, Chow LML, Liu Z (2013) Digital sorting of complex tissues for cell type-specific gene expression profiles. BMC Bioinformatics 14(1):89. https://doi.org/10.1186/1471-2105-14-89
25. Liebner DA, Huang K, Parvin JD (2014) MMAD: microarray microdissection with analysis of differences is a computational tool for deconvoluting cell type-specific contributions from tissue samples. Bioinformatics 30(5):682–689. https://doi.org/10.1093/bioinformatics/btt566
26. Angelova M, Charoentong P, Hackl H, Fischer ML, Snajder R, Krogsdam AM, Waldner MJ, Bindea G, Mlecnik B, Galon J, Trajanoski Z (2015) Characterization of the immunophenotypes and antigenomes of colorectal cancers reveals distinct tumor escape mechanisms and novel targets for immunotherapy. Genome Biol 16(1):64. https://doi.org/10.1186/s13059-015-0620-6
27. Newman AM, Liu CL, Green MR, Gentles AJ, Feng W, Xu Y, Hoang CD, Diehn M, Alizadeh AA (2015) Robust enumeration of cell subsets from tissue expression profiles. Nat Methods 12(5):453–457. https://doi.org/10.1038/nmeth.3337
28. Corliss BA, Azimi MS, Munson JM, Peirce SM, Murfee WL (2016) Macrophages: an inflammatory link between angiogenesis and lymphangiogenesis. Microcirculation 23(2):95–121. https://doi.org/10.1111/micc.12259
29. Aran D, Hu Z, Butte AJ (2017) xCell: digitally portraying the tissue cellular heterogeneity landscape. Genome Biol 18(1):220. https://doi.org/10.1186/s13059-017-1349-1
30. Steen CB, Liu CL, Alizadeh AA, Newman AM (2020) Profiling cell type abundance and expression in bulk tissues with CIBERSORTx.

Methods Mol Biol 2117:135–157. https://doi.org/10.1007/978-1-0716-0301-7_7

31. Newman AM, Steen CB, Liu CL, Gentles AJ, Chaudhuri AA, Scherer F, Khodadoust MS, Esfahani MS, Luca BA, Steiner D, Diehn M, Alizadeh AA (2019) Determining cell type abundance and expression from bulk tissues with digital cytometry. Nat Biotechnol 37(7): 773–782. https://doi.org/10.1038/s41587-019-0114-2
32. Wang J-H, Kumar S, Liu G-S (2021) Bulk gene expression deconvolution reveals infiltration of M2 macrophages in retinal neovascularization. Invest Ophthalmol Vis Sci 62(14):22–22. https://doi.org/10.1167/iovs.62.14.22
33. Barrett T, Wilhite SE, Ledoux P, Evangelista C, Kim IF, Tomashevsky M, Marshall KA, Phillippy KH, Sherman PM, Holko M, Yefanov A, Lee H, Zhang N, Robertson CL, Serova N, Davis S, Soboleva A (2012) NCBI GEO: archive for functional genomics data sets—update. Nucleic Acids Res 41(D1):D991–D995. https://doi.org/10.1093/nar/gks1193
34. Wang JH, Kumar S, Liu GS (2021) Bulk gene expression deconvolution reveals infiltration of M2 macrophages in retinal neovascularization. Invest Ophthalmol Vis Sci 62(14):22. https://doi.org/10.1167/iovs.62.14.22
35. Chen B, Khodadoust MS, Liu CL, Newman AM, Alizadeh AA (2018) Profiling tumor infiltrating immune cells with CIBERSORT. Methods Mol Biol 1711:243–259. https://doi.org/10.1007/978-1-4939-7493-1_12
36. Team RStudio (2021) RStudio: integrated development for R. RStudio, PBC, Boston. 2020

# Chapter 8

# MicroRNA Profiling from Tears as a Potential Non-invasive Method for Early Detection of Diabetic Retinopathy

**Wilson K. M. Wong, Mya Polkamp, Ryan J. Farr, Pooja S. Kunte, Hrishikesh P. Hardikar, Chittaranjan S. Yajnik, Anandwardhan A. Hardikar, and Mugdha V. Joglekar**

## Abstract

Diabetic retinopathy (DR) is a vascular complication of diabetes that can lead to partial or complete loss of vision. Early detection and treatment of DR can prevent blindness. Regular clinical examination is recommended for DR diagnosis; however, it is not always possible or feasible due to limited resources, expertise, time, and infrastructure. Several clinical and molecular biomarkers are proposed for the prediction of DR including microRNAs. MicroRNAs are a class of small non-coding RNAs that are found in biofluids and can be measured using reliable and sensitive methods. The most commonly used biofluid for microRNA profiling is plasma or serum; however, tear fluid (tears) is also demonstrated to contain microRNAs. MicroRNAs isolated from tears present a non-invasive source for DR detection. Different methods of microRNA profiling are available including digital PCR-based methods that can detect up to a single copy of microRNA in the biofluids. Here, we describe microRNA isolation from tears using manual method as well as using a high-throughput automated platform followed by microRNA profiling using digital PCR system.

**Key words** MicroRNAs, Tears, High throughput, Digital PCR, Diabetes, Diabetic retinopathy

## 1 Introduction

Chronic high glucose concentrations in diabetes lead to multiple secondary vascular disorders including diabetic retinopathy (DR), where blood vessels in the retina are damaged causing blurry vision or blindness [1, 2]. Common risk factors of DR involve poor glycemic control, hypertension, diabetes duration, age, high BMI, obesity, and cataract surgery [3, 4]. At present, a variety of clinical features or circulating molecules could be potentially used to predict/diagnose DR. Multiple meta-analyses and systematic reviews have compared the accuracies of different biomarkers for diagnosing DR. One such report found that glycated haemoglobin

Guei-Sheung Liu and Jiang-Hui Wang (eds.), *Diabetic Retinopathy: Methods and Protocols*, Methods in Molecular Biology, vol. 2678, https://doi.org/10.1007/978-1-0716-3255-0_8,

(HbA1c) and 2-h plasma glucose (2h-PG) can be used to diagnose DR [5]. Other biomarkers include retinal photographs/imaging, advanced glycation end products (AGEs), vascular endothelial growth factor (VEGF), cell adhesion molecules (CAMs), and inflammatory markers (cytokines, chemokines) [6, 7]. One of the molecular biomarkers considered for DR prediction is microRNAs (miRNAs) [8].

MiRNAs are ubiquitously expressed, small RNA molecules that do not produce any protein; however, they regulate several cellular functions and physiological processes [9, 10]. They modulate the expression of target genes post-transcriptionally either by inhibiting mRNA translation or by disrupting mRNA stability. Apart from their role in gene regulation, they are excellent candidates for biomarker research. They are highly stable in the bloodstream, are resistant to degradation by freeze-thaw, and can be profiled using sensitive assays [11], making them an attractive blood-based biomarker that can be easily attainable for DR prediction. Several intracellular and circulating miRNAs are being studied and considered as DR biomarkers including but not limited to miR-21, miR-216a, miR-210, miR-221, and miR-151 [12]. In one such study, angiogenic miR-27b and miR-320a were found to be associated with the incidence and progression of DR [13]. Such an array of circulatory miRNAs that display altered levels of expression could be used to identify the increased risk of progression to DR and different stages of DR. A recent systematic review and meta-analysis [14] found statistical evidence in favour of miRNA profiling for human DR.

Apart from plasma and serum, miRNAs have been identified in urine, aqueous and vitreous humor, saliva, and tears [15, 16]. Currently, there is not much known about the miRNA cargo in tears as potential biomarkers of DR progression. Although there have been few studies conducted on miRNAs in tears [17–19], detailed longitudinal studies on changes in miRNA cargo within tears during the development of DR are still lacking. Although DR-specific plasma miRNAs are well-studied, less non-specificity in miRNA profiles is expected from tears, which represent a local milieu of miRNAs from the orbital vasculature.

MiRNAs, either alone or in combination with other clinical examinations or circulating elements, present as important molecular biomarkers of DR. Routine eye check-up is an effective way to monitor vision health in individuals with diabetes, and it can significantly reduce the incidence of DR. Several individuals with diabetes, who are and will need a specialist for an eye check-up, are in areas where ophthalmologists are not readily available (39 ophthalmologists/million in Australia, with 83% of those placed in major cities; 2018 Department of Health AFHW Ophthalmology report). There is a need to address the increasing demand for eye health monitoring in diabetes, where there is no immediate access

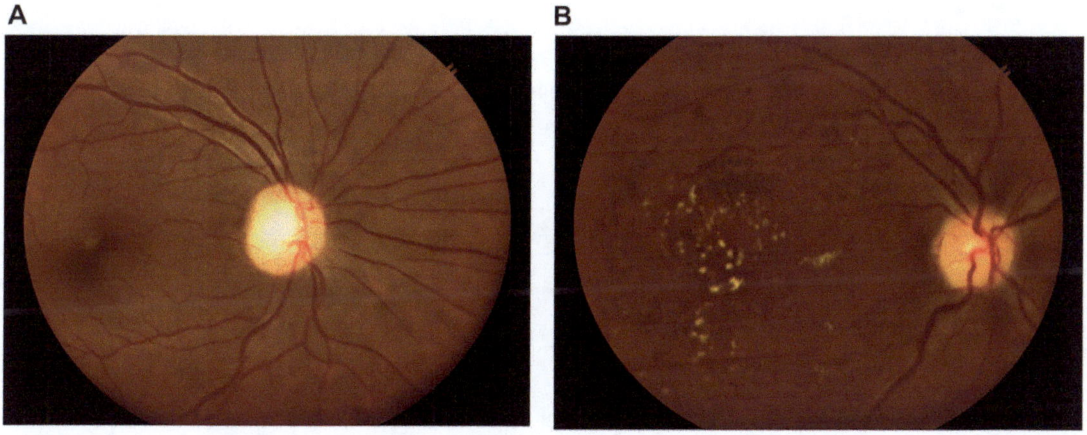


**Fig. 1** Representative images of retina from a healthy person (**a**) and from an individual with diabetes (**b**). Typical "cotton wool" spots are observed in the diabetic retina. Imaging is performed using a digital non-mydriatic fundus camera at the KEM Hospital Research Centre, Pune, India

to specialist care. A non-invasive diagnostic test could help such individuals from remote/regional areas, to monitor the risk of DR in diabetes and prioritize those at the highest risk for ophthalmologist visits and care. Moreover, the current COVID-19 pandemic has underlined the need for a contact-free health monitoring system. A fast, easy, economic, and non-invasive method of DR screening would be highly beneficial.

Non-invasive methods include medical imaging, involving optical coherence tomography (OCT) and retinal fundus photographs (RFP) (Fig. 1) [20]. Easy access to such equipment along with necessary expertise may be a practical challenge in remote communities. Tear miRNA profiling offers another non-invasive method for monitoring DR. Using digital PCR technologies [21, 22], absolute quantitation of miRNAs in tears is also possible. Even though current methods for measuring miRNAs require sophisticated infrastructure and trained personnel, nanotechnology or photonics-based approaches are being developed to circumvent the need for specialized facilities for miRNA profiling. In this chapter, we will detail the methodologies required for RNA isolation, quantitation, and miRNA profiling from tear samples. Combining non-invasive methodologies such as retinal imaging and tear miRNA measurement is a promising way to improve risk prediction of DR in the future (Fig. 2).

## 2 Materials

### 2.1 Tears Collection

1. Schirmer's collection strips.
2. Nuclease-free 1.7 mL microcentrifuge tubes.

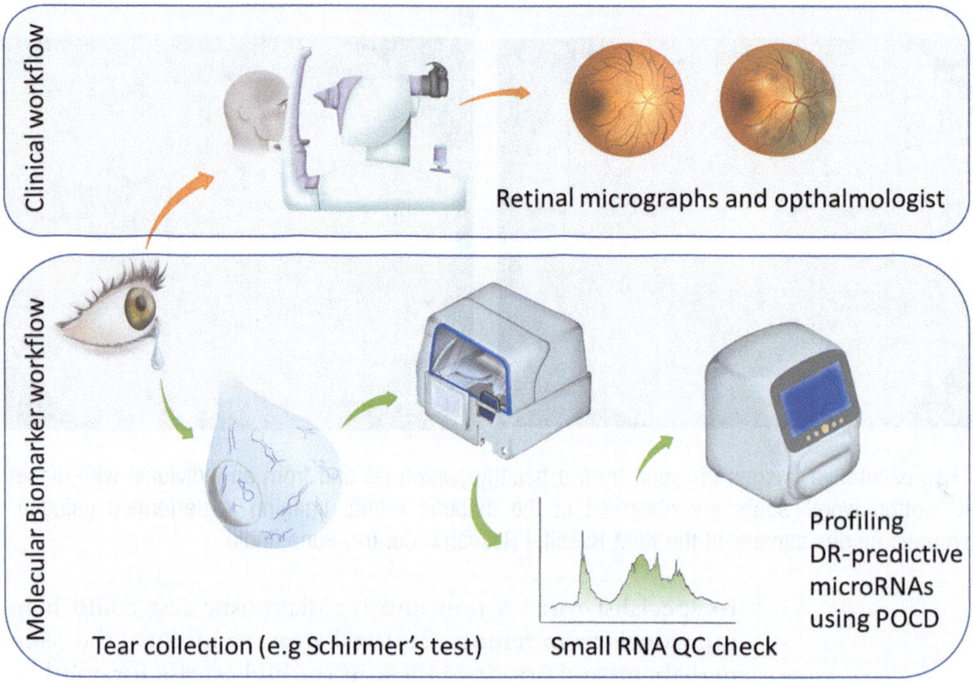


**Fig. 2** A schematic presenting two different non-invasive methodologies for diagnosing DR. Clinical workflow including retinal imaging (top part, orange arrows) is a routinely used method and requires specialist staff to operate and grade the images. Molecular biomarker workflow (bottom part, green arrows) can potentially include tear collection and small RNA isolation and quantification, followed by DR-specific miRNA profiling using point-of-care device (POCD). Combining both non-invasive methodologies in the future will significantly improve risk prediction for diabetic retinopathy

### 2.2 RNA Isolation (Manual Protocol)

1. RNase AWAY® (*see* **Note 1**).
2. Filtered pipette tips.
3. Nuclease-free 1.7 mL microcentrifuge tubes.
4. Glycogen (nuclease-free, molecular biology grade): Working concentration of 5 μg/μL.
5. Synthetic spike-in control (ath-miR-172a: working concentration 50 nM) (*see* **Note 2**).
6. TRIzol©, QIAzol lysis reagent, or TRI reagent (*see* **Note 3**).
7. Chloroform (molecular biology grade).
8. Isopropyl alcohol/IPA (molecular biology grade).
9. Absolute ethanol (96–100% pure, molecular biology grade).
10. Nuclease-free water.

#### 2.3 High-Throughput RNA Isolation on QIAcube HT Robotic Platform

1. RNase AWAY® (*see* **Note 1**).
2. Filtered pipette tips.
3. Collection microtubes (Qiagen) (*see* **Note 4**).
4. Lids/caps for collection microtubes (Qiagen).
5. Glycogen (nuclease-free, molecular biology grade): Working concentration of 10 μg/μL.
6. Synthetic spike-in control (ath-miR-172a: working concentration 50 nM) (*see* **Note 2**).
7. TRIzol©, QIAzol lysis reagent, or TRI reagent (*see* **Note 3**).
8. Chloroform (molecular biology grade).
9. Isopropyl alcohol/IPA (molecular biology grade).
10. Absolute ethanol (96–100% pure, molecular biology grade).
11. Nuclease-free water.
12. RNeasy 96 QIAcube HT kit (Qiagen).
13. QIAcube HT Plasticware (Qiagen).
14. Tip disposal box (Qiagen).
15. Reagent trough with lid, 7 mL (Qiagen).
16. Reagent trough with lid, 170 mL (Qiagen).
17. Filter tips, OnCor C, 200 μL (Qiagen).
18. TriGene (or any other cleaning detergent).

#### 2.4 Tear-Derived Small RNA Quantification on Agilent Bioanalyzer

1. RNase AWAY® (*see* **Note 1**).
2. Agilent Small RNA Kit that includes small RNA chips, small RNA ladder, small RNA dye concentrate, small RNA marker, small RNA gel matrix, safe-lock 0.5 mL microcentrifuge tubes (DNase-/RNase-free), syringe, spin filters, and 2× electrode cleaner (Agilent Technologies) (*see* **Note 5**).
3. Nuclease-free 0.2 mL microcentrifuge tubes.
4. Nuclease-free water.

#### 2.5 MicroRNA Reverse Transcription

1. TaqMan™ MicroRNA Reverse Transcription (RT) Kit containing 10× RT buffer, dNTP mix with dTTP (100 mM), RNase inhibitor (20 U/μL), and MultiScribe™ reverse transcriptase enzyme (50 U/μL).
2. Synthetic spike-in control (ath-miR-159a: working concentration 15 nM) (*see* **Note 2**).
3. Mature miRNA-specific reverse transcription (RT) primers (these will vary with the target of interest).
4. Nuclease-free water.

### 2.6 MicroRNA Quantification Using Digital PCR

1. QIAcuity Probe PCR Kit (Qiagen) (*see* **Note 6**).
2. QIAcuity Nanoplates and Accessories containing QIAcuity 26k or 8.5k 24-well Nanoplates (Qiagen).
3. Nuclease-free 0.2 mL microcentrifuge tube(s) (*see* **Note 7**).
4. TaqMan® PCR primer-probe assay (20×) for the specific miRNA target of interest, fluorescence dye (e.g., FAM) labelled. It is important to use the RT and PCR primers for the same miRNA.

## 3 Methods

Use RNase AWAY® as appropriate. Always handle the samples, tubes, and consumables including Schirmer's strips with gloves to avoid any contamination or degradation. All reagents used are molecular biology grade and all plasticware nuclease-free.

### 3.1 Tears Collection

Schirmer's strip contains a line of dye and a graduated measurement (in mm). As tears are collected, the dye runs along with the tears to indicate how much tear is collected (in mm) on the strip.

1. Aliquot 200 μL of TRIzol into 1.7 mL microcentrifuge tube (s) (*see* **Note 8**).
2. Bend Schirmer's strip over just below the dye line, and place the strip near the lower eyelid.
3. Let the tears soak up on Schirmer's strip. Record the distance travelled by the dye on the strip.
4. Collect tears following **steps 1** and **2** (Subheading 3.1), separately from each individual or from each eye, as per the experimental plan/research question.
5. Transfer Schirmer's strip (with tears) into a 1.7 mL tube containing the 200 μL of TRIzol. Another Schirmer's strip without any tears is placed into a different tube with 200 μL of TRIzol as a control.
6. Vortex the tube to allow mixing of the tears on Schirmer's strip with TRIzol (for 30 s), and then spin down the tube on a minifuge at maximum speed for 10 s.
7. Transfer the TRIzol mixed with tears into a new 1.7 mL microcentrifuge tube. These samples can be processed immediately for RNA isolation or can be stored at −80 °C for use in the future after proper labelling. Discard the tube containing Schirmer's strip.

### 3.2 RNA Isolation (Manual Method)

RNA isolation can be performed manually if there are fewer samples; however, for larger clinical studies, we recommend using an automated platform (detailed in Subheading 3.3) to maintain consistency across multiple samples/batches and save significant time and labor costs. Below are the steps for the manual protocol for RNA isolation.

1. Thaw the tear samples (collected in 200 μL of TRIzol) on ice if they have been stored at −80 °C.
2. Vortex and briefly spin glycogen (5 μg/μL), and then add 1 μL to each sample.
3. Add 2.5 μL of 50 nM ath-miR-172a spike in control to each tube.
4. Add 40 μL of chloroform to each tube, and manually shake the tubes vigorously for 40 s (*see* **Note 8**). The tubes can be placed in a rack and secured properly with your hands before shaking. This ensures proper mixing of TRIzol and chloroform after addition. Do not vortex.
5. Incubate at room temperature for 15 min.
6. Centrifuge 12,000 × *g* for 15 min at 4 °C.
7. Transfer the upper aqueous (clear) phase to a fresh 1.7 mL microcentrifuge tube. Do not disturb the interface (cloudy precipitate layer). Try to remove as much of the aqueous layer without contamination from the layers below.
8. Add 100 μL of IPA to each tube (that contains the clear aqueous layer).
9. Invert each tube for five to ten times.
10. Incubate at room temperature for 10 min.
11. Centrifuge at 12,000 × *g* for 8 min at 4 °C. Always orient the tubes with the hinge facing the outer circumference of the centrifuge. This method helps in locating the RNA pellet, which will be at the bottom toward the hinge after centrifugation. This is important when the starting material, such as tears, has lower RNA content.
12. Prepare a fresh volume of 75% ethanol using absolute ethanol and nuclease-free water, while samples are spinning.
13. Remove and discard the supernatant by placing the pipette tip along the wall of the tube opposite to the hinge. When removing the supernatant, start with a larger-sized pipette tip (P200) to a smaller-sized tip (P10) when getting closer to the pellet. A pellet may not always be visible; if this is the case, assume the pellet is sitting on side of the wall which has the hinge.
14. Add 200 μL of freshly prepared 75% ethanol to each tube and vortex three times. The pellet may be seen floating in the solution.

15. Centrifuge at 7500 × *g* for 5 min at 4 °C.
16. Remove and discard supernatant carefully as described in **step 13** (Subheading 3.2).
17. Air-dry the tubes at room temperature for 5 min. If there are large droplets, gently twist the tube to spread out the liquid, allowing it to dry faster. Do not overdry the samples as this will reduce the solubility of the RNA.
18. Elute each pellet in 10–15 μL of nuclease-free water, resuspend, vortex, and spin down. Always store RNA on ice after this stage. The elution volume may vary depending on the pellet size.
19. Store RNA at −80 °C.

### 3.3 High-Throughput RNA Isolation on QIAcube HT Robotic Platform

Below are the steps for performing RNA isolation on a QIAcube HT automated platform that can process a minimum of 24 to a maximum of 96 samples at one time.

1. Thaw the tear samples (collected in 200 μL of TRIzol) on ice if they have been stored at −80 °C.
2. Make up the volume to 500 μL with fresh TRIzol in a fume hood (*see* **Note 8**).
3. Add 100 μL of nuclease-free water to the required number of collection microtubes (*see* **Note 9**).
4. Add the 500 μL of the sample (tears in TRIzol) to separate collection microtubes, and mix with nuclease-free water (100 μL, already added to each collection microtube) by gently pipetting up and down.
5. Record the position of the samples added to the collection microtubes. Ensure that a control sample (without tears) is included in the isolation process.
6. Vortex and briefly spin glycogen (10 μg/μL stock), and then add 1 μL to each sample.
7. Add 2.5 μL of 50 nM ath-miR-172a spike in control to each tube.
8. Add 100 μL of chloroform to each tube; close the tubes using specialized lids/caps suitable for the collection microtubes. Cover the capped collection microtubes with a plastic lid.
9. Shake vigorously for 40 s. Do not vortex (*see* **Note 8**).
10. Incubate at room temperature for 15 min.
11. Centrifuge at 3200 × *g* for 25 min at 4 °C.
12. Turn on the QIAcube HT system. Remove the tip eject cover (red). Open the pre-treatment run file (*see* **Note 10**).
13. Place collection microtubes into position C1 of the QIAcube HT. The trough holder in this position must be removed first.

14. Load an empty S-block into position B1. Load the required number of tips.
15. Begin the run. This program will transfer 300 μL of aqueous phase into the S-block.
16. Remove the collection microtubes. If needed, one can store the phenolic fractions remaining in the collection microtubes at −80 °C for protein/DNA analysis. Place the trough holder back into this position.
17. Open the purification run file (*see* **Note 10**).
18. Load the appropriate reagents in their respective troughs. See software for specific volumes.
19. Load RNeasy plate and elution plate in their respective positions. Ensure that well A1 of both these plates is at the top left corner.
20. Load tips (2 × 96). Start the run and complete the pre-run checklist.
21. Remove the elution plate after completion of the run. Cap and label these tubes of the elution plate and then store at −80 °C.
22. Clean QIAcube HT system by removing the reagent troughs and discarding the remaining liquid.
23. Remove the channel block (three pieces), and pour 20 mL of RO water down the waste chute. Replace the tip eject cover (red).
24. Soak channel block in 1% TriGene for 15–30 min. Rinse with RO water. Dry.
25. Run the cleaning cycle, including UV, if this is the last run of the day. Select to turn off after the cleaning cycle is completed.

### 3.4 Tear-Derived Small RNA Quantification on Agilent Bioanalyzer

RNA quantification can be performed using various methods. The simplest method includes a NanoDrop measurement that provides the concentration. The Agilent bioanalyzer analyzes RNA quantity as well as quality. It is important to remember that the addition of synthetic miRNA spike-in control would affect the quantity and electropherogram. To obtain an unaltered quantity and distribution of miRNAs from tears, it is advised to perform RNA isolation without spike-in control.

1. Thaw the Agilent Small RNA Kit at room temperature for at least 1 h or at 25 °C for 10 min before use. Keep small RNA kit reagents away from light exposure, as they are light sensitive.
2. Thaw the RNA samples on ice if frozen.
3. Filter the gel, by pipetting the entire volume (~650 μL) of a small RNA gene matrix into a fresh spin filter (from the kit).
4. Centrifuge at 10,000 × *g* for 15 min at room temperature.

5. Aliquot 40 μL of the filtered gel into a fresh 0.5 mL RNase-free microcentrifuge tube (from the kit). Once filtered, the remaining unused gel is stored at 4 °C and should be used within 4 weeks.
6. Vigorously vortex (for 10 s) and spin down the dye concentrate (from the kit).
7. Add 2 μL of the dye into the 40 μL filtered gel aliquot.
8. Mix gel-dye mix by pipetting, inverting, and flicking the tube five to ten times.
9. Centrifuge gel-dye mix at 13,000 × *g* for 10 min at room temperature. The prepared gel-dye mix must be used within 1 day and kept away from light.
10. Prepare μL aliquots of small RNA ladder (from the kit and stored in −20 °C) into fresh 0.2 mL microcentrifuge tubes. Keep one aliquot for immediate use, and store the other aliquots in a −80 °C freezer for future use.
11. Aliquot 2 μL of each of the tear RNA samples in fresh, separate 0.2 mL microcentrifuge tubes.
12. Denature RNA samples and the small RNA ladder by incubating at 70 °C for 2 min. This can be achieved in the thermocycler with the lid and heat block set at 70 °C.
13. Place denatured RNA samples and the small RNA ladder immediately on ice for at least 1 min.
14. Add 350 μL of RNase AWAY to wash chip #1 (supplied in the kit and is reusable), and place it onto the electrodes of the device with the device lid closed for 1 min. This process cleans the bioanalyzer electrodes prior to the actual sample run. Label this wash chip #1 as RNase AWAY wash chip for future use.
15. Open lid, remove RNase AWAY wash chip from the device, and discard RNase AWAY solution and rinse with 350 μL of fresh RNase AWAY.
16. Rinse another wash chip (wash chip #2, from the kit) with 350 μL of a nuclease-free water. Label this wash chip #2 as nuclease-free water wash chip for future use.
17. Add fresh 400 μL of nuclease-free water to the nuclease-free water wash chip, and place it onto a device with device lid closed for 5 min to now wash the device electrodes with nuclease-free water.
18. Remove the nuclease-free water wash chip from the device, discard nuclease-free water, and rinse with fresh 400 μL of nuclease-free water.
19. Repeat **steps 17** and **18** (Subheading 3.4).

20. Open the lid of the device to allow the electrodes to air-dry for 1 min, and then close the lid.
21. Set up the priming station with a syringe from the kit attached and with the settings for a small RNA chip (the base set at "C" and the lever set to the bottom position).
22. Place a new small RNA chip (from the kit) onto the priming station.
23. Load 9 μL the gel-dye mix onto the allocated well marked with a white-colored "G" (*see* **Note 11**).
24. Set the plunger to the 1 mL mark of the syringe, and then close the chip priming station (until a click sound is heard).
25. Press the plunger down until it is held by the lever's clip.
26. Wait for 60 s (exactly) and then release the lever.
27. Wait for 5 s to allow the plunger to come back up, and then slowly pull back the plunger to the 1 mL mark (if it has not reached back to the 1 mL mark).
28. Open the chip priming station, and load 9 μL of the gel-dye mix in each of the two black-colored "G"-marked wells on the chip.
29. Spin down the conditioning solution (from the kit) at the maximum speed on the minifuge for 10 s, and load 9 μL of conditioning solution into the well-marked "CS."
30. Vortex the small RNA marker vigorously, and spin down at maximum speed on the minifuge for 10 s (supplied in the kit).
31. Load 5 μL of the small RNA marker into the remaining 12 wells of the chip.
32. Load 1 μL of denatured small RNA ladder into the allocated well marked with a ladder symbol on the chip.
33. Load 1 μL of the denatured tear RNA sample into their allocated wells on the chip. Add 1 μL of the small RNA marker to any unused sample well(s). Record the loading position of the samples on the chip.
34. Place the chip into the IKA vortex (*see* **Note 12**), and vortex the chip at max speed for 1 min. Make sure the chip is secure on the vortex.
35. Open the bioanalyzer software program, and set it up for small RNA chip analysis.
36. Place a small RNA chip onto the bioanalyzer and close the lid.
37. Click "Run," and let the bioanalyzer analyze the samples on the small RNA chip. The run should take up to 45 min. The run should be started within 5 min after vortexing the chip.
38. Analyze the output (Fig. 3a) using Microsoft Excel or GraphPad Prism (Fig. 3b) (*see* **Note 13**).

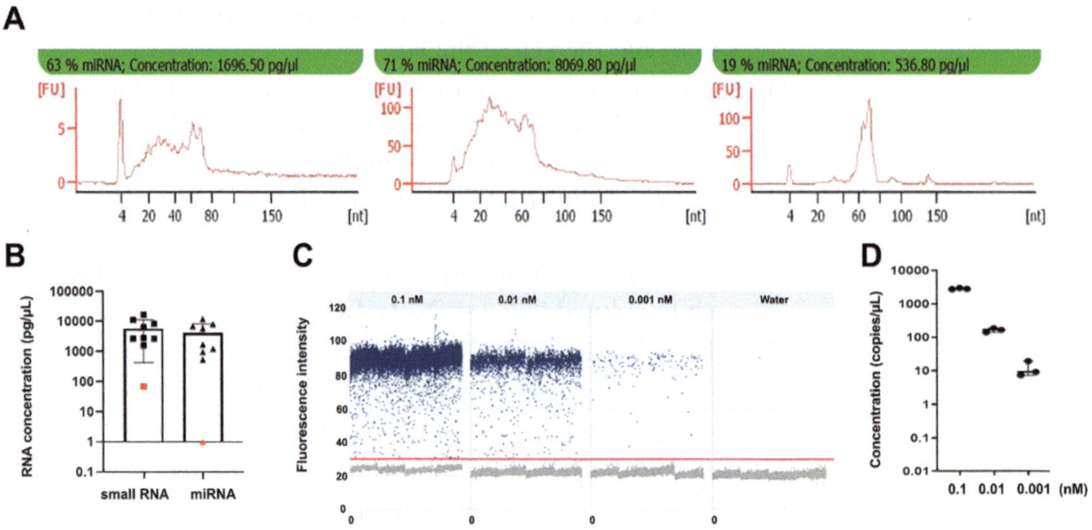


**Fig. 3** (**a**) Representative examples of the small RNA chip bioanalyzer electropherogram results for three different tear samples (unaltered/without any miRNA spike-in control added). The RNA regions are determined by the read-outs of the small RNA ladder peaks. The size and concentration of a sample are calculated based on the x-axis (migration time) and y-axis (fluorescence intensity (FU)) compared to the reference ladder. (**b**) The concentration of unaltered total small RNA and miRNA from tear samples (in pg/μL) is presented here. Data were obtained as part of a preliminary and feasibility study on $n = 5$ individuals (after consent) and are presented as bar graphs (mean with SD) where each point is a different reading from an individual (one/two readings for each participant). Red color points indicate data from no tear/control sample. (**c**) Example of the one-dimensional (1D) plot produced after digital PCR on the QIAcuity platform. The fluorescence intensity (*y*-axis) of each partition droplet for each well (*x*-axis) is presented. The threshold (red line across the plot) can be set manually for each or multiple wells. (**d**) Representative data from a QIAcuity digital PCR experiment using different dilutions (*x*-axis) of a spike-in miRNA. The miRNA concentration (in copies/μL) as obtained from the QIAcuity system is presented on the *y*-axis. Three replicates of each dilution were used in the experiment

### 3.5 MicroRNA Reverse Transcription with Spike-In Control

1. Thaw all reagents except the reverse transcriptase enzyme and RNA (if frozen) on ice (*see* **Note 14**).
2. Gently vortex all reagents (except the reverse transcriptase enzyme), and briefly spin them down at maximum speed on the minifuge for 10 s.
3. Dilute the RNA if necessary to around 10 ng/μL, as each miRNA reverse transcription reaction would require 10 ng of total RNA in a total of 5 μL total reaction volume.
4. Aliquot 10 ng of each RNA sample into an individual 0.2 mL microcentrifuge tube, and add an appropriate volume of nuclease-free water to make a total of 1.67 μL. Keep the diluted RNA on ice. If you are assessing more than one miRNA for each sample, then prepare the required number of tubes for each sample and each with 10 ng of RNA in 1.67 μL of volume.

**Table 1**
**Calculations of reagents required for a single microRNA reverse transcription mix**

| Reagent | Volume per reaction (μL) | Volume per 96 samples (including 5% excess) (μL) |
|---|---|---|
| 10× RT buffer | 0.5 | 50.4 |
| 100 mM dNTPs | 0.05 | 5.04 |
| RNase inhibitor | 0.06 | 6.05 |
| Reverse transcriptase enzyme | 0.33 | 33.26 |
| ath-miR-159a spike-in | 0.1 | 10.08 |
| Nuclease-free water | 1.29 | 130.03 |
| *Total* | 2.33 | 234.86 |

5. To prepare the reverse transcription master mix for one microRNA/sample, combine 0.5 μL of 10× RT buffer, 0.05 μL of 100 mM dNTPs, 0.06 μL of RNase inhibitor, 0.33 μL of reverse transcriptase enzyme, 0.1 μL ath-miR-159a spike-in control (15 nM), and 1.29 μL of nuclease-free water for a total volume of 2.33 μL (Table 1). Calculate for and prepare at least 5% excess volume to compensate for any pipetting error (*see* **Note 15**).
6. Add 2.33 μL of the reverse transcription master mix into each 0.2 mL microcentrifuge tube containing diluted RNA.
7. Add 1 μL of miRNA reverse transcription primer to each sample tube, to make a final reaction volume of 5 μL (*see* **Note 16**).
8. Invert the 0.2 mL microcentrifuge tubes three to four times to mix, and briefly centrifuge at maximum speed on the minifuge for 10 s.
9. Place the tubes into a thermocycler with the following cycling conditions, 16 °C for 30 min, 42 °C for 30 min, and 85 °C for 5 min, and hold at 4 °C.
10. Store cDNA at −20 °C or use immediately for downstream digital PCR.

### 3.6 Digital PCR MicroRNA Target Quantification

1. Place the cDNA (if frozen), PCR primer-probe (20×) for specific target miRNA of interest (has to be for the same miRNA used in reverse transcription step), and QIAcuity Probe PCR Kit components (Probe-PCR master mix (4×) and RNase-free water) on wet ice for thawing.
2. Vortex Probe-PCR master mix and PCR primer-probe before use.

**Table 2**
**Calculations of reagents required for dPCR master mix**

| Component | For one reaction in the 26k Nanoplate (μL) | For one reaction in the 8.5k Nanoplate (μL) |
|---|---|---|
| Probe-PCR master mix (4×) | 10 | 3 |
| Primer-probe assay (20×) | 2 | 0.6 |
| Nuclease-free water | 26 | 7.4 |
| Total volume | 38 | 11 |

3. Prepare the digital PCR reaction mix by adding 10 μL of Probe-PCR master mix (4×), 2 μL of PCR primer-probe assay (20×), and 26 μL of nuclease-free water for a total volume of 38 μL for one reaction in the 26k Nanoplate (Table 2). These volumes are for one sample; calculate and prepare reaction mix for the required number of samples with an extra half (sample) reaction to avoid any errors. There are different types of QIAcuity Nanoplates (number of nano chambers/partitions, 26k or 8.5k, and number of wells per Nanoplate, 24-well and 96-well). Prepare the reaction mix based on the number of samples and the type of Nanoplate (*see* **Notes 17–19**).
4. Vortex the prepared reaction mix.
5. Load the appropriate volume of reaction mix (in this example, 38 μL per reaction for 26k and 11 μL per reaction for 8.5k Nanoplate) into newly allocated 0.2 mL microcentrifuge tubes.
6. Vortex and spin down the thawed cDNA samples. Load 2 μL (for 26k Nanoplate preparation) or 1 μL (for 8.5k Nanoplate preparation) of the cDNA sample into the respective 0.2 mL microcentrifuge tubes (containing the reaction mix). Add nuclease-free water (in place of cDNA) in one of the tubes as a no template control.
7. Resuspend 3 to 5 times or vortex gently the reaction mix and cDNA for complete mixing (while avoiding air bubble generation), and then transfer the entire volume of the reaction from the 0.2 mL microcentrifuge tube onto the bottom of each allocated well in the Nanoplate. Direct addition to the bottom of each well and using only the first stop of the pipette during the transfer will help in avoiding any air bubbles at this step (*see* **Note 20**).
8. Seal the Nanoplate with Nanoplate seal (supplied in the Nanoplate accessories kit) by carefully placing the sticky side onto one edge of the plate covering the plate entirely. Avoid having the seal cover over the side edges of the plate (*see* **Note 21**).

9. Use the roller (provided with the QIAcuity platform) to seal the sides and across the plate to make sure the plate is sealed properly.
10. Turn on the QIAcuity instrument and software.
11. Set up the run by entering the primer-probe assay details used and the digital PCR condition, as per the QIAcuity manual: PCR initial heat activation for 2 min at 95 °C for 1 time and two-step cycling (denaturation for 15 s at 95 °C and combined annealing/extension for 30 s at 60 °C) for 40 times.
12. Place the plate into the QIAcuity platform, and start the digital PCR run following instructions in the experimental software.
13. Assess the results and, as necessary, set the threshold based on the control (no template control) used and re-analyze (with the new threshold set) the results (Fig. 3c, d).

## 4 Notes

1. RNase AWAY is highly recommended to be used as a cleaning reagent for gloves and benchtops to help remove RNA-degrading enzymes.
2. An alternate synthetic miRNA spike-in control (not known to be expressed in your samples) can be designed and used. Spike-in controls are specifically useful for normalizing the miRNA data from biofluids.
3. We do not see differences in the RNA yields with these different TRIzol reagents obtained from different companies.
4. This protocol is provided for QIAcube HT for high-throughput RNA isolation. All reagents and consumables recommended and essential for this workflow are to be purchased from Qiagen. There are other automated and high-throughput workflows for RNA isolation from other companies, and in that case, one should use the recommended kits and consumables for the specific equipment available to them.
5. This protocol is written for RNA quality control check using Agilent bioanalyzer, and therefore, reagents and consumables required for this instrument are provided in the "Materials" section. If one has access to other equipment that can perform the same function, they will have to use the reagents and kits for those specialized systems.
6. This protocol provides workflow of digital PCR using QIAcuity digital PCR system from Qiagen. Necessary reagents and consumables are listed with recommended providers. There are other digital PCR platforms as outlined in our previous work [21, 22], and one must use reagents/consumables that are compatible with their system.

7. One can use 0.2 mL 8-well strips or 96-well plates as per their experimental design.
8. Handle TRIzol, chloroform, isopropyl alcohol, and ethanol in the fume hood. Always wear appropriate PPE and safety goggles while handling TRIzol.
9. This addition is to ensure appropriate volumes at the time of using automated pipette handling systems. One may have to optimize the volumes as per the instrument available in their lab.
10. Pre-treatment and purification are customized protocols for the use of QIAcube HT to transfer the aqueous phase and extract RNA, respectively. These were developed by the Qiagen team for our laboratory.
11. When loading gel-dye mix, small RNA marker, condition solution, ladder, and samples into the allocated wells in the small RNA chip in this (**step 23** of Subheading 3.4) and later steps (**steps 28–33** of Subheading 3.4), make sure to load the solution directly into the bottom of the well and avoid producing air bubbles. These air bubbles will interfere with the results.
12. A specialized vortex that can hold the small RNA chip is required at this stage.
13. One can use other available RNA chips to understand the total RNA quality and quantity in tear samples.
14. The reverse transcriptase enzyme should always be stored at −20°C and removed only on a −20 °C cold block just before addition. Reverse transcriptase enzymes should not be vortexed and should only be spun down at maximum speed on a minifuge for 10 s before addition.
15. The calculations are for one miRNA and one sample. You can increase the volumes of these reagents based on the number of miRNAs and the number of samples planned in your experiment.
16. Perform all additions on ice and always keep the reaction tubes on ice.
17. The volume of RNase-free water may vary depending (i) upon the concentration (thereby volume) of the primer/probe mix used, (ii) if multiple (duplex or multiplex) primer-probes are used in a single reaction, and (iii) upon the volume of cDNA sample.
18. This protocol provides details of using a TaqMan™ PCR primer-probe assay for miRNA quantification on the QIAcuity digital PCR platform. Hence, Probe-PCR master mix details are included. One can also use EvaGreen master mix for primer sets (forward and reverse primers, without probes).

19. Different primer(s), primer-probes of interest for miRNAs of interest with FAM, VIC (HEX), TAMRA, ROX, or Cy5 fluorophore dye, can be used.
20. Air bubbles will lead to inaccurate loading, fewer number of reactions, and incorrect measurement of the gene expression.
21. When sealing the plate, the seal should avoid covering the side edge of the plate. If the seal covers the edge sides of the plate, it may interfere with the run on the QIAcuity platform. To avoid having the seal cover the edge of the plate, one can turn the plate with the bottom side of the plate facing towards you (as bottom side of the plate has more space between the wells and edge of plate) and peel the seal sticker halfway and slowly place seal onto the left edge of the plate to help guide the alignment of the seal to the plate.

## Acknowledgments

We would like to thank Dr. Gerald Liew for providing Schirmer's strips and all participants who collected tear samples as well as consented for medical imaging for this study.

### References

1. Wang W, Lo ACY (2018) Diabetic retinopathy: pathophysiology and treatments. Int J Mol Sci 19(6). https://doi.org/10.3390/ijms19061816
2. Vujosevic S, Aldington SJ, Silva P, Hernandez C, Scanlon P, Peto T, Simo R (2020) Screening for diabetic retinopathy: new perspectives and challenges. Lancet Diabetes Endocrinol 8(4):337–347. https://doi.org/10.1016/S2213-8587(19)30411-5
3. Ting DS, Cheung GC, Wong TY (2016) Diabetic retinopathy: global prevalence, major risk factors, screening practices and public health challenges: a review. Clin Exp Ophthalmol 44(4):260–277. https://doi.org/10.1111/ceo.12696
4. Magliah SF, Bardisi W, Al Attah M, Khorsheed MM (2018) The prevalence and risk factors of diabetic retinopathy in selected primary care centers during the 3-year screening intervals. J Family Med Prim Care 7(5):975–981. https://doi.org/10.4103/jfmpc.jfmpc_85_18
5. Martinez-Vizcaino V, Cavero-Redondo I, Alvarez-Bueno C, Rodriguez-Artalejo F (2016) The accuracy of diagnostic methods for diabetic retinopathy: a systematic review and meta-analysis. PLoS One 11(4): e0154411. https://doi.org/10.1371/journal.pone.0154411
6. Jenkins AJ, Joglekar MV, Hardikar AA, Keech AC, O'Neal DN, Januszewski AS (2015) Biomarkers in diabetic retinopathy. Rev Diabet Stud 12(1–2):159–195. https://doi.org/10.1900/RDS.2015.12.159
7. Pusparajah P, Lee LH, Abdul Kadir K (2016) Molecular markers of diabetic retinopathy: potential screening tool of the future? Front Physiol 7:200. https://doi.org/10.3389/fphys.2016.00200
8. Joglekar MV, Januszewski AS, Jenkins AJ, Hardikar AA (2016) Circulating microRNA biomarkers of diabetic retinopathy. Diabetes 65(1):22–24. https://doi.org/10.2337/dbi15-0028
9. Ambros V (2004) The functions of animal microRNAs. Nature 431(7006):350–355. https://doi.org/10.1038/nature02871
10. Dexheimer PJ, Cochella L (2020) MicroRNAs: from mechanism to organism. Front Cell Dev Biol 8:409. https://doi.org/10.3389/fcell.2020.00409
11. Farr RJ, Januszewski AS, Joglekar MV, Liang H, McAulley AK, Hewitt AW, Thomas HE, Loudovaris T, Kay TW, Jenkins A,

Hardikar AA (2015) A comparative analysis of high-throughput platforms for validation of a circulating microRNA signature in diabetic retinopathy. Sci Rep 5:10375. https://doi.org/10.1038/srep10375
12. Smit-McBride Z, Morse LS (2021) MicroRNA and diabetic retinopathy-biomarkers and novel therapeutics. Ann Transl Med 9(15):1280. https://doi.org/10.21037/atm-20-5189
13. Zampetaki A, Willeit P, Burr S, Yin X, Langley SR, Kiechl S, Klein R, Rossing P, Chaturvedi N, Mayr M (2016) Angiogenic microRNAs linked to incidence and progression of diabetic retinopathy in type 1 diabetes. Diabetes 65(1):216–227. https://doi.org/10.2337/db15-0389
14. Ma L, Wen Y, Li Z, Wu N, Wang Q (2022) Circulating microRNAs as potential diagnostic biomarkers for diabetic retinopathy: a meta-analysis. Front Endocrinol (Lausanne) 13:929924. https://doi.org/10.3389/fendo.2022.929924
15. Weber JA, Baxter DH, Zhang S, Huang DY, Huang KH, Lee MJ, Galas DJ, Wang K (2010) The microRNA spectrum in 12 body fluids. Clin Chem 56(11):1733–1741. https://doi.org/10.1373/clinchem.2010.147405
16. Cortez MA, Bueso-Ramos C, Ferdin J, Lopez-Berestein G, Sood AK, Calin GA (2011) MicroRNAs in body fluids--the mix of hormones and biomarkers. Nat Rev Clin Oncol 8(8):467–477. https://doi.org/10.1038/nrclinonc.2011.76
17. Chan HW, Yang B, Wong W, Blakeley P, Seah I, Tan QSW, Wang H, Bhargava M, Lin HA, Chai CH, Mangunkusumo EA, Thet N, Yuen YS, Sethi R, Wang S, Hunziker W, Lingam G, Su X (2020) A pilot study on microRNA profile in tear fluid to predict response to anti-VEGF treatments for diabetic macular edema. J Clin Med 9(9). https://doi.org/10.3390/jcm9092920
18. Pinazo-Duran MD, Zanon-Moreno V, Lleo-Perez A, Garcia-Medina JJ, Galbis-Estrada C, Roig-Revert MJ, Marco-Ramirez C, Lopez-Galvez M, Dolz-Marco R, Duarte L, Campos Borges C, Salgado-Borges J, Gallego-Pinazo R (2016) Genetic systems for a new approach to risk of progression of diabetic retinopathy. Arch Soc Esp Oftalmol 91(5):209–216. https://doi.org/10.1016/j.oftal.2016.01.016
19. Sun L, Liu X, Zuo Z (2021) Regulatory role of miRNA-23a in diabetic retinopathy. Exp Ther Med 22(6):1477. https://doi.org/10.3892/etm.2021.10912
20. Salz DA, Witkin AJ (2015) Imaging in diabetic retinopathy. Middle East Afr J Ophthalmol 22(2):145–150. https://doi.org/10.4103/0974-9233.151887
21. Maynard CL, Wong WKM, Hardikar AA, Joglekar MV (2019) Droplet digital PCR for measuring absolute copies of gene transcripts in human islet-derived progenitor cells. Methods Mol Biol 2029:37–48. https://doi.org/10.1007/978-1-4939-9631-5_4
22. Farr RJ, Wong WKM, Maynard CL, Tersey SA, Mirmira RG, Hardikar AA, Joglekar MV (2019) Comparative analysis of diagnostic platforms for measurement of differentially methylated insulin DNA. J Biol Methods 6(2). https://doi.org/10.14440/jbm.2019.280

# Chapter 9

# Proximity Extension Assay (PEA) Platform to Detect Vitreous Biomarkers of Diabetic Retinopathy

**Ricardo Lamy, Showgy Ma'ayeh, Sarantis Chlamydas, and Jay M. Stewart**

## Abstract

Diabetic retinopathy (DR) is one of the leading causes of blindness, affecting more than 100 million people worldwide. Currently, DR prognosis and management are based mainly on biomarkers identified by direct retinal fundus observation or by imaging devices. The use of molecular biology to discover biomarkers of DR has great potential to impact the standard of care, and the vitreous humor can serve as an indirect source for those molecular biomarkers because it is rich in proteins secreted by the retina. Proximity extension assay (PEA) is a technology that combines antibody-based immunoassays with DNA-coupled methodology to obtain information on the abundance of multiple proteins while using minimal sample volume, with high specificity and sensitivity. Matched antibodies labelled with a complementary sequence of oligonucleotides are used to simultaneously bind a target protein in solution, and when in proximity, the complementary sequences on each antibody hybridize, serving as template for DNA polymerase-dependent extension and the generation of a unique double-stranded DNA "barcode." PEA works well with vitreous matrix and has great potential to support the identification of novel predictive and prognostic biomarkers of DR.

**Key words** Diabetes, Diabetic retinopathy, Vitreous biomarkers, Vitreous proteomics, Liquid biopsy, Molecular diagnosis, Multiplex assay, Drug discovery

## 1 Introduction

Diabetic retinopathy (DR) is one of the leading causes of preventable blindness and the most common microvascular complication of diabetes mellitus (DM), affecting about 22% of the DM population [1]. DM is characterized by hyperglycemia secondary to inadequate insulin production and/or resistance to insulin action [2]. Sustained hyperglycemia can lead to ocular neurovascular abnormalities and DR through a series of mechanisms and molecular pathways, such as oxidative stress, increased generation of advanced glycation end products, and overactivation of the polyol and protein kinase C pathways (Fig. 1) [3].

Guei-Sheung Liu and Jiang-Hui Wang (eds.), *Diabetic Retinopathy: Methods and Protocols*, Methods in Molecular Biology, vol. 2678, https://doi.org/10.1007/978-1-0716-3255-0_9,

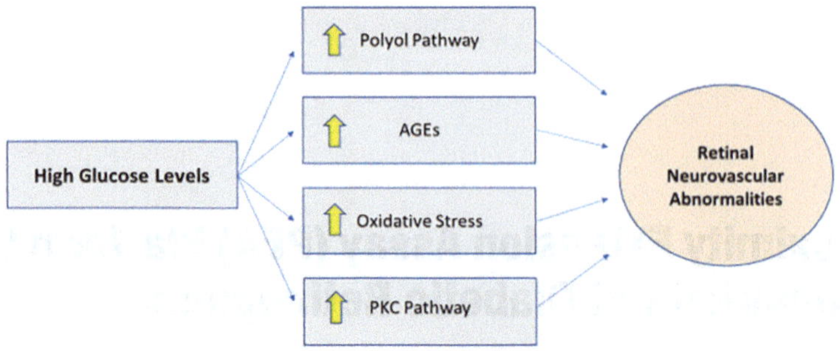


**Fig. 1** Schematic presentation of potential mechanisms and pathways involved in the development of retinal neurovascular abnormalities in diabetic retinopathy (DR). Sustained high levels of glucose can lead to overactivation of the polyol pathway, hyperactivation of the protein kinase C (PKC) pathway, and an increase in the production of advanced glycation end products (AGEs) and reactive oxygen species. (Reproduced from Lamy R. and Stewart J.M. 2022 with permission from Springer-Nature)

Many retinal structural changes and signs of DR can be observed with an ophthalmoscopic exam. DR stratification, treatment, and medical decisions are based mainly on imaging biomarkers that are commonly obtained from photos of the fundus of the eye and from optical coherence tomography [2, 4, 5]. Molecular biomarkers (i.e., non-imaging biomarkers that are measurable based on their molecular properties, including nucleic acids and proteins) may provide powerful additional information about DR and can potentially be detected before the manifestation of any structural change [6–8]. Blood and plasma are the most common sources of molecular biomarkers in medicine, but they do not always correlate well with ocular diseases. Due to the small volume of the eye in relation to the body, local variations in the abundance of some proteins are difficult to detect systemically, and therefore, ocular specimens are usually a better source for researching novel biomarkers of retinal diseases. The risk of complications associated with retinal biopsies does not justify its clinical use for detecting biomarkers of DR; however, liquid biopsies of the adjacent tissues (i.e., vitreous and aqueous) can be safely obtained without damaging the neurovascular structure of the retina [9–11].

The vitreous is a gel composed mainly of water, collagen, and hyaluronan that fills the posterior chamber of the eye and is in close contact with the retina [12]. It contains several molecules released by the retinal cells and vasculature that can be correlated with the physiologic status of the retina [13]. Recent advancements in proteomics techniques, including the development of novel multiplexing platforms, have enabled assaying several proteins using small sample volume, revolutionizing the study of molecular biomarkers in ocular specimens [14, 15]. Recently, Lamy et al. have shown the efficacy of using the PEA technology to detect several proinflammatory and angiogenic cytokines in the vitreous humor of patients

**Table 2**
**PCR protocol used in the extension and amplification steps of the proximity extension assay**

| Step | Temperature (°C) | Duration |
|---|---|---|
| Extension | 50 | 20 min |
| Hot start | 95 | 5 min |
| PCR cycle (17 cycles) | 95 | 30 s |
| | 54 | 1 min |
| | 60 | 1 min |
| Maintain reaction at | 10 | ∞, hold |

**Table 3**
**Extension mixture recipe (sufficient for a 96-well plate)**

| Extension mixture | Volume (μL) |
|---|---|
| PEA enzyme[a] | 55 |
| PEA polymerase[a] | 22 |
| PEA solution[a] | 1100 |
| High-purity water | 9385 |
| **Total** | **10,562** |

*PEA* proximity extension assay
[a]Olink® Target 96 Panel Kit

(d) Spin down the incubation plate at 400–1000×*g* for 1 min at room temperature.

(e) Keep PEA and PCR enzymes in an ice block until their addition to the extension reaction mixture.

To prepare the extension reaction mixture (enough for a 96-well plate), use the volumes presented in Table 3. Add the enzymes last. Use the centrifuge tubes (>11 mL) to prepare the extension reaction mixture.

2. The Extension Reaction

(a) Vortex the mixture prepared above, and pour into a multichannel pipette reservoir (time-sensitive step).

(b) Start a timer for 5 min, and pipette 96 μL of the extension mix into the upper part of the well in the incubation plate (reverse pipetting).

(c) Seal the plate with adhesive plastic film, vortex thoroughly (using MixMate®), and spin down.

(d) Place the plate in the thermal cycler and start the PEA program (Table 2).

### 3.4 Detection Step

The final step in the PEA protocol is the detection step which quantifies the DNA barcodes for each biomarker using high-throughput real-time qPCR (e.g., Olink® Signature Q100 system or the Fluidigm® BioMark system).

This step starts with priming the Dynamic Array IFC that has two accumulators. Each accumulator is injected with one control line fluid syringe. It is important to ensure that each valve is open before continuing with the procedure. The IFC chip is then primed on the qPCR machine (*see* **Note 12**).

1. Detection Reaction

   (a) Remove the incubation plate (with extension products) from the PCR instrument, vortex, and spin down. Inspect the plate, and note any variations in the volume of liquid in each well (*see* **Note 13**). Set aside for **step 5**.

   (b) Thaw the primer plate and the detection solution, vortex, and spin down (*see* **Note 14**).

   (c) Prepare the detection mixture (sufficient for a 96-well plate) in a microfuge tube (1–1.5 mL as detailed in Table 4). Keep the enzyme in ice blocks until their addition to the mixture.

   (d) Vortex the detection mix, spin briefly, and pipette 95 μL into each well of an eight-well strip. Transfer 7.2 μL from the strip to the wells of new 96-well plate (reverse pipetting). Mark the plate as the "sample plate."

   (e) Transfer 2.8 μL from each well in the incubation plate to the corresponding well in the sample plate (forward pipetting). Seal plate with adhesive film, vortex, and spin down.

**Table 4**
**Detection mixture recipe (sufficient for a 96-well plate)**

| Detection mixture | Volume (μL) |
|---|---|
| Detection solution[a] | 550 |
| PCR polymerase[a] | 3.1 |
| Detection enzyme[a] | 7.8 |
| High-purity water | 230 |
| **Total** | **790.9** |

*PCR* polymerase chain reaction
[a]Olink® Target 96 Panel Kit

(f) Transfer 5 μL from each well of the primer plate and 5 μL of the sample plate into the primed 96.96 IFC left and right inlets, respectively (reverse pipetting) (*see* **Notes 15–17**).

(g) Load the IFC on the qPCR machine, following the instructions on the screen.

(h) Analyze your results using the Olink® NPX manager software. Data requires normalization by inter-plate normalization (one plate) or intensity normalization (multiple plate project) prior to conducting further analyses.

## 4 Notes

1. Always perform PEA assays in a clean laboratory environment. Wear a lab coat and gloves while performing the procedure.
2. Set up the lab space to have separate pre-PCR and post-PCR rooms. Do not move consumables and equipment between different labs.
3. Clean the lab bench and all surfaces (ventilation hood, pipettes, racks) in the working area with 70% ethanol before starting the procedure.
4. Ensure that you have all the materials needed in the protocol prior to starting.
5. Ensure that pipettes used in this protocol are always well maintained and calibrated. Maintain and calibrate lab equipment regularly (e.g., PCR machines and BioMark™).
6. Distribute samples into an 8-well strip or a 96-well plate prior to starting to reduce sample handling time.
7. A multichannel pipette and reverse pipetting must be used in the reagent transfer step.
8. To maximize accuracy, use reagents and liquids only after they have reached room temperature, and pre-rinse the tips with the liquid to be pipetted one to three times before reverse pipetting.
9. Inspect the plate when vortexing to ensure even mixing of the liquids in all wells.
10. Ensure that all wells in the 96-well plates are sealed properly.
11. For the incubation steps, samples should be pipetted into a 96-well plate or 8-well strips. Vortexing should be done using MixMate® (or similar) at 2500 rpm for 30 s, and centrifugations are done at 400–1000 × *g* for 1 min at room temperature.

12. Prepare the detection mix while the IFC is priming.
13. The incubation plate, with extension products, can be stored for up to 1 week at +4 °C or up to 4 weeks at −20 °C in a refrigerator or freezer with the light off when the door is closed.
14. Remove the adhesive film from the primer plate carefully to avoid contamination between wells.
15. The IFC has two parts: the part to the left is for loading the primers and the part to the right for samples. A correct IFC orientation can be verified when the cut corner of the chip is on the upper left side. The IFC should be loaded carefully to avoid cross-contamination of primers or samples. Load the first column of the sample plate (A1-H1), followed by the second column (A2-H2) starting from the second well on the IFC. See the pattern of loading (Fig. 4). The same is done for the primer plate.
16. Change tips when transferring both primers or samples from the plate to the IFC.
17. Do not leave any inlets on the IFC empty. Inspect the plate and remove any bubbles that will interfere with the run.

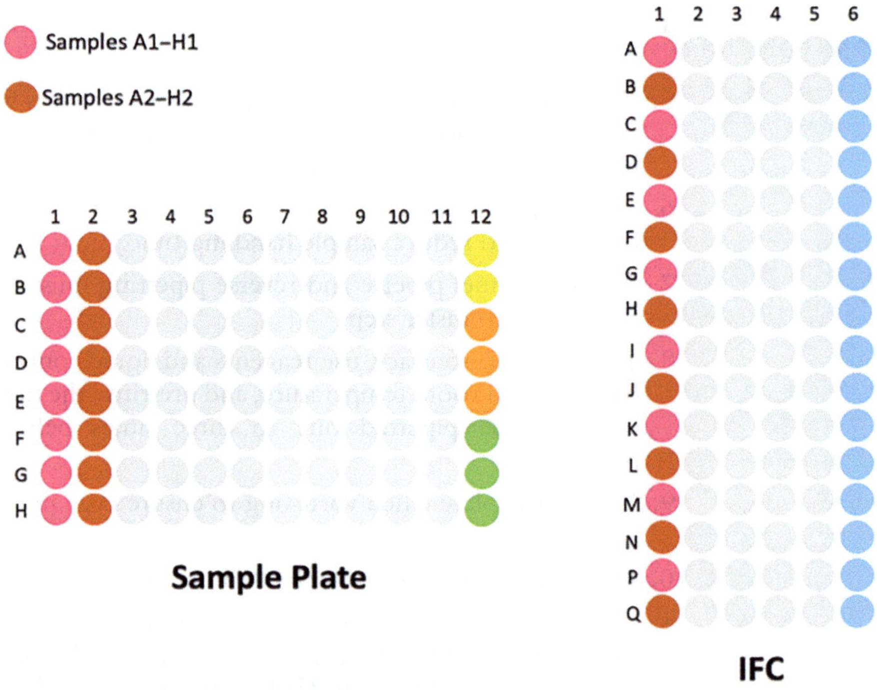


**Fig. 4** Illustration of the IFC chip template used in the detection step

## References


1. Teo ZL, Tham YC, Yu M, Chee ML, Rim TH, Cheung N, Bikbov MM, Wang YX, Tang Y, Lu Y, Wong IY, Ting DSW, Tan GSW, Jonas JB, Sabanayagam C, Wong TY, Cheng CY (2021) Global prevalence of diabetic retinopathy and projection of burden through 2045: systematic review and meta-analysis. Ophthalmology. https://doi.org/10.1016/j.ophtha.2021.04.027
2. American Diabetes Association (2021) 2. Classification and diagnosis of diabetes: Standards of Medical Care in Diabetes-2021. Diabetes Care 44(Suppl 1):S15–S33. https://doi.org/10.2337/dc21-S002
3. Lamy R, Stewart JM (2022) Vitreous biomarkers: what they are and how they may be used to advance the management of diabetic retinopathy. In: Patel VB, Preedy VR (eds) Biomarkers in diabetes, Biomarkers in disease: methods, discoveries and applications. Springer, Cham. https://doi.org/10.1007/978-3-030-81303-1_51-1
4. Flaxel CJ, Adelman RA, Bailey ST, Fawzi A, Lim JI, Vemulakonda GA, Ying GS (2020) Diabetic retinopathy preferred practice pattern(R). Ophthalmology 127(1):P66–P145. https://doi.org/10.1016/j.ophtha.2019.09.025
5. Wilkinson CP, Ferris FL 3rd, Klein RE, Lee PP, Agardh CD, Davis M, Dills D, Kampik A, Pararajasegaram R, Verdaguer JT, Global Diabetic Retinopathy Project Group (2003) Proposed international clinical diabetic retinopathy and diabetic macular edema disease severity scales. Ophthalmology 110(9):1677–1682. https://doi.org/10.1016/S0161-6420(03)00475-5
6. Broza YY, Zhou X, Yuan M, Qu D, Zheng Y, Vishinkin R, Khatib M, Wu W, Haick H (2019) Disease detection with molecular biomarkers: from chemistry of body fluids to nature-inspired chemical sensors. Chem Rev 119(22):11761–11817. https://doi.org/10.1021/acs.chemrev.9b00437
7. Laterza OF, Hendrickson RC, Wagner JA (2007) Molecular biomarkers. Ther Innov Regul Sci 41:573–585. https://doi.org/10.1177/009286150704100504
8. FDA-NIH Biomarker Working Group (2020) BEST (biomarkers, EndpointS, and other tools). Food and Drug Administration (US) 2016-. Contents of a Biomarker Description, Silver Spring
9. Midena E, Frizziero L, Midena G, Pilotto E (2021) Intraocular fluid biomarkers (liquid biopsy) in human diabetic retinopathy. Graefes Arch Clin Exp Ophthalmol 259(12):3549–3560. https://doi.org/10.1007/s00417-021-05285-y
10. Skeie JM, Brown EN, Martinez HD, Russell SR, Birkholz ES, Folk JC, Boldt HC, Gehrs KM, Stone EM, Wright ME, Mahajan VB (2012) Proteomic analysis of vitreous biopsy techniques. Retina 32(10):2141–2149. https://doi.org/10.1097/IAE.0b013e3182562017
11. Velez G, Tang PH, Cabral T, Cho GY, Machlab DA, Tsang SH, Bassuk AG, Mahajan VB (2018) Personalized proteomics for precision health: identifying biomarkers of vitreoretinal disease. Transl Vis Sci Technol 7(5):12. https://doi.org/10.1167/tvst.7.5.12
12. Bishop PN (2000) Structural macromolecules and supramolecular organisation of the vitreous gel. Prog Retin Eye Res 19(3):323–344. https://doi.org/10.1016/s1350-9462(99)00016-6
13. de Smet MD, Gad Elkareem AM, Zwinderman AH (2013) The vitreous, the retinal interface in ocular health and disease. Ophthalmologica 230(4):165–178. https://doi.org/10.1159/000353447
14. Assarsson E, Lundberg M, Holmquist G, Bjorkesten J, Thorsen SB, Ekman D, Eriksson A, Rennel Dickens E, Ohlsson S, Edfeldt G, Andersson AC, Lindstedt P, Stenvang J, Gullberg M, Fredriksson S (2014) Homogenous 96-plex PEA immunoassay exhibiting high sensitivity, specificity, and excellent scalability. PLoS One 9(4):e95192. https://doi.org/10.1371/journal.pone.0095192
15. Lundberg M, Eriksson A, Tran B, Assarsson E, Fredriksson S (2011) Homogeneous antibody-based proximity extension assays provide sensitive and specific detection of low-abundant proteins in human blood. Nucleic Acids Res 39(15):e102. https://doi.org/10.1093/nar/gkr424
16. Lamy R, Farber-Katz S, Vives F, Ayanoglu G, Zhao T, Chen Y, Laotaweerungsawat S, Ma D, Phone A, Psaras C, Li NX, Sutradhar S, Carrington PE, Stewart JM (2020) Comparative analysis of multiplex platforms for detecting vitreous biomarkers in diabetic retinopathy. Transl Vis Sci Technol 9(10):3. https://doi.org/10.1167/tvst.9.10.3
17. Olink (2022) Olink Target 96 user manual. https://www.olink.com/content/uploads/2022/04/olink-target-96-usermanual.pdf. Accessed 28 Jul 2022

# Chapter 10

# Methods for Analyzing the Gut Microbiota in Diabetic Retinopathy

**Yinhua Huang, Jacey Hongjie Ma, Yanfang Peng, Xu Yang, Jiansu Chen, and Shibo Tang**


## Abstract

The gut microbiome that inhabits human hosts plays an important role in the development of a healthy host immune system. Many studies have shown that gut microbiota is involved in the occurrence and development of diabetic retinopathy (DR). With the advent of sequencing technology of the bacterial 16S ribosomal RNA (rRNA) gene, microbiota studies are becoming more feasible. Here, we described a study protocol to characterize the microbiota composite in the DR and non-DR patients compared with healthy controls.


**Key words** Gut microbiota, 16s, Diabetic retinopathy

## 1 Introduction

Diabetic retinopathy (DR) is a sight-threatening microvascular complication in diabetic patients, which is a leading cause of blindness and vision impairment in the working-age population [1]. Approximately 50% of people with type 2 diabetes may develop retinopathy. Despite significant research, the mechanisms underpinning its pathogenesis remain poorly understood. The major risk factors for developing DR include hypertension, hyperglycemia, and diabetes duration, but which of these abnormalities contribute to recognized features of DR is still unknown. It is important to note that despite these systemic risk factors, DR does not occur in some people with diabetes, and glycemic control may not be necessary to eliminate the lifetime risk of DR [2]. These patterns indicate that other factors, such as genetic susceptibility and environment, are involved in the initiation and progression of DR. Therefore, it is often difficult to predict the risk of DR in individuals with diabetes.

Guei-Sheung Liu and Jiang-Hui Wang (eds.), *Diabetic Retinopathy: Methods and Protocols*, Methods in Molecular Biology, vol. 2678, https://doi.org/10.1007/978-1-0716-3255-0_10,

The innumerable number of gut microbiota colonizes the human intestine. The bacterial number in the adult gut reaches $10^{14}$, weighing around 1.13 and 10 times more than the total number of human cells [3]. They constitute the normal gut microbiota which compose a symbiotic complex and quiet active ecosystem. These microorganisms interact with the external and host environment, which influences many physiological and pathophysiological processes, such as nutrient production, signaling pathway modulation, energy homeostasis, and production of anti-inflammatory responses [4].

Many studies have shown that gut microbiome dysbiosis occurs in many diseases. However, the composition of the gut microbiota varies in different people and various diseases. A disturbance of the gut microbiome, such as the decrease of beneficial bacteria and the emergence of harmful bacteria, is common in patients with prediabetes and type 2 diabetes mellitus (T2DM) [5]. Our previous studies have found several gut microbiota dysbiosis in patients with DR and DM compared with healthy individuals [6], which suggest that the characteristic gut microbiota may serve as a biomarker to diagnose or predict DR.

Microbes inhabit the intestinal tract in large numbers. However, it is difficult to recognize and define the entire flora in the human gut, as many members of the gut microbiota have been refractory to culture using traditional microbiological techniques. Despite the advances in the culture method over the past 20 years, it is impossible to culture all members of the gut microbiome. In the 1980s, the development of polymerase chain reaction (PCR) revolutionized the field of biomedical research [7]. Recent technical advances have allowed researchers to study gut microbiota using high-throughput sequencing techniques rather than traditional culture techniques. Independent amplification of conserved genes or gene regions in all prokaryotes, followed by sequencing of these PCR products, can provide a detailed overview of the taxa present in a given sample, such as a fecal sample. Although similar methods have applied to fungi and protozoa, this chapter focuses on characterizing bacterial communities in human fecal samples.

## 2 Materials

### 2.1 Fecal Sample Collection and Storage

1. Toilet paper.
2. Sterile scoop and stool containers.
3. Ice box.
4. Liquid nitrogen.

### 2.2 DNA Extraction

1. E.Z.N.A.® Soil DNA Kit.
2. NanoDrop 2000 UV-vis spectrophotometer.
3. 1% agarose gel.

### 2.3 PCR Amplification and Fluorescence Quantitation

1. TransStart FastPfu DNA Polymerase.
2. AxyPrep DNA Gel Extraction Kit.
3. 2% agarose gel.
4. QuantiFluor™-ST blue fluorescence quantification system.

### 2.4 MiSeq Library Construction and Sequencing

1. DNA Library Prep kit.
2. Illumina MiSeq platform.

## 3 Methods

### 3.1 Study Subject Recruitment and Fecal Sample Collection

1. The exclusion criteria of the subjects
   - (a) Have a history of cancer or other intestinal inflammatory diseases.
   - (b) Take major gastrointestinal surgery such as cholecystectomy, appendectomy, etc. within 5 years.
   - (c) Invasive medical intervention within 3 months prior to sampling.
   - (d) Take antibiotics, probiotics, or prebiotics within 4 weeks before sample collection (*see* **Note 1**).
   - (e) Diet changed significantly in the week prior to sampling.
   - (f) Menstruating woman.
2. Collection and preservation of fecal samples
   - (a) Exhaust urine: ensure urine be excreted before collecting stool samples, to avoid urine contamination of feces.
   - (b) Defecate: toilet paper can be matted in the urinal and defecate on it.
   - (c) Sampling: wash hands clean, and scoop out the part inside the stool that does not touch the air and ground with a sterile scoop; scoop out two scoopfuls (3–5 g).
   - (d) The excavated feces were collected into sterile feces containers and transferred to the laboratory within two hours (*see* **Note 2**).
   - (e) After being sub-packaged, snap-frozen fecal samples were placed in liquid nitrogen and immediately stored at −80 °C until analysis.

### 3.2 Sequence Data Acquisition

1. DNA Extraction

   Microbial DNA was extracted from frozen fecal samples using E.Z.N.A.® Soil DNA Kit according to the manufacturer's protocol. DNA purity and concentration were determined using a NanoDrop 2000 UV-vis spectrophotometer, and DNA integrity was checked via 1% agarose gel electrophoresis.

2. PCR Amplification

   The V3 and V4 hypervariable regions of the bacterial 16s rRNA gene were chosen as targets for amplification using PCR with indexed barcodes. The primer sequences used were 338F (5′-ACTCCTACGGGAGGCAGCAG-3′) and 806R (5′-GGA CTACHVGGGTWTCTAAT -3′). PCR reactions were performed using TransStart FastPfu DNA Polymerase in 20 μL reaction mixtures. The amplicon products were purified using an AxyPrep DNA Gel Extraction Kit. The PCR products from three repeats of a sample were mixed and detected by 2% agarose gel electrophoresis.

3. Fluorescence Quantitation

   Purified amplicon products were determined by QuantiFluor™-ST blue fluorescence quantification system based on the preliminary quantitative results of electrophoresis. After that, the corresponding proportion of mixing was carried out according to the sequencing amount requirements of each sample.

4. MiSeq Library Construction

   Steps for constructing the library were as follows:

   (a) Connecting the "Y"-shaped connector.
   (b) Using magnetic beads to screen and remove self-connecting fragments in the connector.
   (c) Using PCR amplification to enrich the library templates.
   (d) Denaturing with sodium hydroxide to produce single-stranded DNA fragments.

5. Illumina Sequencing.

   The sequencing process is based on the MiSeq platform (Illumina). Briefly steps for MiSeq sequencing were as follows:

   (a) One end of the DNA fragment is complementary to the primer base and fixed on the chip.
   (b) The other end is randomly complementary with another primer nearby, which is also fixed to form a "bridge."
   (c) PCR amplification, producing DNA clusters.
   (d) The DNA amplicon linearizes into a single strand.
   (e) The modified DNA polymerase and dNTP with four fluorescent markers were added to synthesize only one base per cycle.

(f) The surface of the reaction plate was scanned with a laser, and the nucleotide species polymerized in the first reaction of each template sequence were read.

(g) The "fluorophorer" and "terminator" are chemically cleaved to restore the viscosity of the 3′-end and continue to polymerize the second nucleotide.

(h) The results of fluorescence signals collected in each round were counted, and the sequence of template DNA fragments was obtained.

### 3.3 Sequence Data Analysis

Sequence data was a gene set generated from 16s rRNA gene sequence and sequenced as paired-end reads on Illumina platform. Sequence data was required in FASTQ format. Every sample has a FASTQ sequence file, and all of the FASTQ files were placed into the same directory. PE reads obtained by MiSeq sequencing were firstly spliced according to overlap relationship, and the sequence quality was controlled and filtered at the same time. OTU cluster analysis and species taxonomy analysis were performed after the samples were distinguished. A typical analysis pipeline is shown in Fig. 1.

1. Data optimization

   To obtain clean data, the raw sequence data were merged, and quality control was performed using Trimmomatic and then merged by adjusting short reads in several steps.

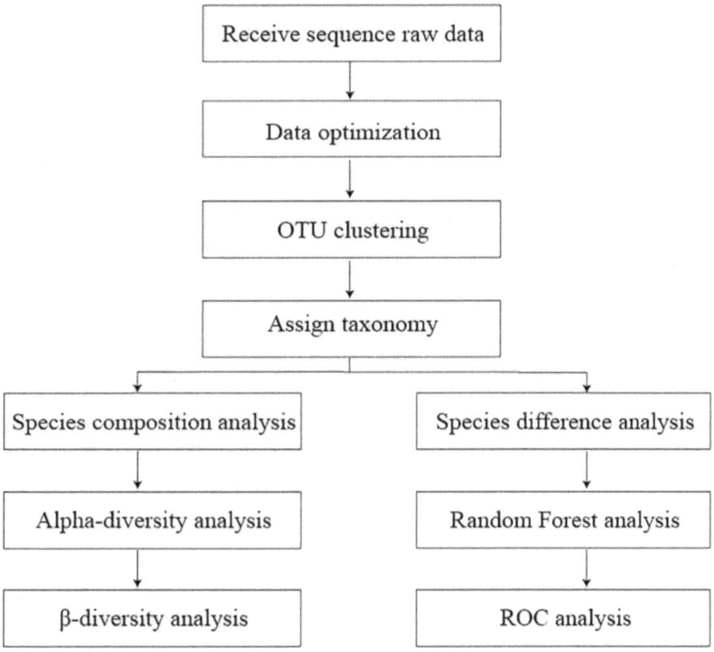


**Fig. 1** Flowchart showing the main steps in the sequence data analysis

(a) A 50 bp window was set for filtering reads with quality scores lower than 20. If the average mass value in the window was less than 20, the back-end bases were cut off from the window, and reads with a tail mass value of less than 50 bp were filtered after quality control, and the reads containing N bases were removed.

(b) According to the correlation between PE reads, pairs of reads were merged into a sequence with a minimum length of overlap of 10 bp.

(c) Nonconforming sequences were removed with a mismatch ratio of no more than 0.2.

(d) Samples were separated according to the barcode and primer at both ends of the sequence, and the sequence direction was adjusted. The allowed mismatch number of barcodes was 0, and the maximum primer mismatch number was 2.

2. OTU clustering

The high-quality sequences were clustered into operational taxonomic units (OTUs) with 97% similarity cutoffs using UPARSE v7.1 (http://drive5.com/uparse/).

(a) Optimized sequences were used to extract non-repetitive ones to reduce redundant computations.

(b) Remove single sequences that are not repeated.

(c) OTU clustering was performed for non-repeating sequences (excluding single sequences) according to 97% similarity. Chimeras were removed during the clustering process to obtain representative sequences of OTU.

(d) All optimized sequences were mapped and selected to OTU representative sequences with more than 97% similarity in order to generate OTU tables.

3. Assign taxonomy

The taxonomy of each 16S rRNA gene sequence was analyzed using the Ribosomal Database Project (RDP) Classifier v2.11 (https://sourceforge.net/projects/rdp-classifier/) based on the Silva database v132 (https://www.arb-silva.de/). The RDP Classifier Bayesian algorithm was used for the taxonomic analysis of OTU representative sequences at 97% similar level. The community composition was analyzed at each classification level, including domain, kingdom, phylum, class, order, family, genus, and species.

4. Species composition analysis

By comparing one or more samples at different classification levels, it can be known what common and unique microbes were present and the relative abundance of individual microbes in the sample. Usually, 97% OTU or other taxonomic sample tables are used for analysis.

5. Alpha-diversity analysis

Alpha diversity refers to the diversity within a particular region or ecosystem, reflecting the richness, evenness, diversity, and coverage of microbial communities.

(a) Community richness was commonly assessed by computing the species indices (sobs, chao, ace, jack, bootstrap).

(b) Community evenness was commonly assessed by computing the species indices (simpsoneven, shannoneven, heip).

(c) Community diversity was commonly assessed by computing the species indices (shannon, simpson, invsipson, qstst).

(d) Community coverage was commonly assessed by computing the species indices (coverage).

6. Beta-diversity analysis

(a) Principal component analysis (PCA)

The PCA diagram reflects the differences of multiple sets of data on the two-dimensional coordinate graph using variance decomposition, and the coordinate axis takes the two characteristic values that can best reflect the variance value. For example, the differences and distances between samples can be reflected by analyzing OTU composition of different samples; the more similar the sample composition, the closer the distance reflected in the PCA diagram.

(b) Principal coordinate analysis (PCoA)

PCoA analysis is a non-constrained data dimension reduction analysis method, similar to PCA analysis. The main difference is that PCA is based on Euclidean distance, while PCoA is not limited to the distance algorithm. The calculation is equivalent to a rotation of the sample distance matrix, which does not change the reciprocal positions of sample points, but changes the coordinate system. After the raw OTU table was normalized, it was calculated with Bray-Curtis, unweight and weight UniFrac distance, using a permutation test to detect differences among the three groups.

(c) Partial least squares discriminant analysis (PLS-DA)

PLS-DA is a supervised analysis method, which requires the detection samples to be grouped according to their categories. In the calculation of the mathematical model, the groups are distinguished, the random differences within the groups are ignored, and the systematic differences between the groups are highlighted. When the observation factor is much larger than the number of samples, it has a good sample discrimination performance. Permutational multivariate analysis of variance was used to test differences in the three groups and was performed in R Studio v3.7.1 using the package "vegan," with argument permutations = 999 and distance = "Bray-curtis."

7. Species difference analysis

   The species between different groups (or samples) of microbial communities were tested for hypothesis, and the significance level of species abundance differences was evaluated to obtain the species with significant differences between groups (or samples). Linear discriminant analysis effect size (LEfSe) analysis mainly consists of three steps: First, the non-parametric Kruskal-Wallis sum-rank test was used to detect species abundance differences between different groups and obtain significantly different species. Second, the Wilcoxon rank-sum test was used to test the consistency of differences among subgroups of different species. Third, an LDA linear discriminant analysis was used to estimate the influence of these different species on differences between groups (*see* **Note 3**). LEfSe was carried out to identify the different bacterial taxa in the groups. Finally, the differential species were used as biomarkers for further study. To identify discriminate microbial markers across the DR, DM, and HC groups, the different OTUs or other classification levels among the three groups were analyzed using LEfSe analysis. Then, all discriminative OTUs or other classification levels were input for the random forest classifier (Python's scikit-learn package) to predict the discrimination between the group of DR/DM, DR/HC, and DM/HC.

8. Random Forest and ROC analysis

   To acquire the best discriminant performance of taxa across patient varieties, the relative abundances of bacterial taxa in the phylum, class, order, family, genus, and OTU levels against different groups can be classified using the random forest classifier (R, randomForest package). For example, we used family taxa set and the random forest function to generate the classification model to predict the discrimination in DM vs. HC, DR vs. HC, and DM vs. DR. In each combination, 500 trees were considered. The receiver operating characteristic (ROC) curve was obtained for the display of the constructed models to evaluate the biomarkers for the random forest, and then the area under the curve (AUC) was used to designate the ROC effect (R, plotROC package) [8].

## 4 Notes

1. Avoid antibiotics use before fecal samples are collected as antibiotics may significantly affect the intestinal microbiota [9].
2. Avoid fecal sample contamination. Store samples at a low temperature as soon as the samples are obtained.
3. In the LEfSe analysis result, LDA score of significant difference is generally set at 2. The LDA threshold could be increased if there are too many obtained species.

### References

1. Tarr JM, Kaul K, Chopra M, Kohner EM, Chibber R (2013) Pathophysiology of diabetic retinopathy. ISRN Ophthalmol 2013:343560. https://doi.org/10.1155/2013/343560
2. Wong TY, Sun J, Kawasaki R, Ruamviboonsuk P, Gupta N, Lansingh VC, Maia M, Mathenge W, Moreker S, Muqit MMK, Resnikoff S, Verdaguer J, Zhao P, Ferris F, Aiello LP, Taylor HR (2018) Guidelines on diabetic eye care: the International Council of Ophthalmology recommendations for screening, follow-up, referral, and treatment based on resource settings. Ophthalmology 125(10):1608–1622. https://doi.org/10.1016/j.ophtha.2018.04.007
3. Shivaji S (2017) We are not alone: a case for the human microbiome in extra intestinal diseases. Gut Pathog 9:13. https://doi.org/10.1186/s13099-017-0163-3
4. Pflughoeft KJ, Versalovic J (2012) Human microbiome in health and disease. Annu Rev Pathol 7:99–122. https://doi.org/10.1146/annurev-pathol-011811-132421
5. Wang J, Ma Q, Li Y, Li P, Wang M, Wang T, Wang C, Wang T, Zhao B (2020) Research progress on traditional Chinese medicine syndromes of diabetes mellitus. Biomed Pharmacother 121:109565. https://doi.org/10.1016/j.biopha.2019.109565
6. Huang Y, Wang Z, Ma H, Ji S, Chen Z, Cui Z, Chen J, Tang S (2021) Dysbiosis and implication of the gut microbiota in diabetic retinopathy. Front Cell Infect Microbiol 11:646348. https://doi.org/10.3389/fcimb.2021.646348
7. Saiki RK, Scharf S, Faloona F, Mullis KB, Horn GT, Erlich HA, Arnheim N (1985) Enzymatic amplification of beta-globin genomic sequences and restriction site analysis for diagnosis of sickle cell anemia. Science 230(4732):1350–1354. https://doi.org/10.1126/science.2999980
8. Zheng P, Yang J, Li Y, Wu J, Liang W, Yin B, Tan X, Huang Y, Chai T, Zhang H, Duan J, Zhou J, Sun Z, Chen X, Marwari S, Lai J, Huang T, Du Y, Zhang P, Perry SW, Wong ML, Licinio J, Hu S, Xie P, Wang G (2020) Gut microbial signatures can discriminate unipolar from bipolar depression. Adv Sci (Weinh) 7(7):1902862. https://doi.org/10.1002/advs.201902862
9. Dethlefsen L, Relman DA (2011) Incomplete recovery and individualized responses of the human distal gut microbiota to repeated antibiotic perturbation. Proc Natl Acad Sci U S A 108 (Suppl 1):4554–4561. https://doi.org/10.1073/pnas.1000087107

# Chapter 11

# Mass Spectrometry-Based Vitreous Proteomics: Validated Methods and Analysis Pipeline


**Sarah Weber, Nick Carruthers, Chris Gates, Yuanjun Zhao, and Jeffrey Sundstrom**



## Abstract

Retinal diseases like diabetic retinopathy and age-related macular degeneration affect millions of individuals worldwide and often lead to vision loss. Vitreous fluid abuts the retina, is accessible for sampling, and contains many proteins related to retinal disease. Therefore, analysis of vitreous is an important tool for studying retinal disease. Because it is rich in proteins and extracellular vesicles, mass spectrometry-based proteomics is an excellent method for vitreous analysis. Here, we discuss important variables to consider when performing vitreous proteomics via mass spectrometry.



**Key words** Retinal disease, Vitreous, Mass spectrometry, Proteomics, Extracellular vesicles


## 1 Introduction

Retinal diseases like diabetic retinopathy (DR) and age-related macular degeneration (AMD) affect millions of individuals worldwide and are among the leading causes of vision loss. Despite their prevalence, therapeutic options to slow or prevent disease progression are limited. This is in part due to the inability to directly biopsy retinal tissue for molecular analysis to guide treatment. However, numerous studies in the past several years have demonstrated that vitreous humor contains proteins related to retinal disease [1–5] and is easily and safely accessible for sampling [6]. The richness of information within vitreous may be due to its abundance of extracellular vesicles (EVs) (*see* **Note 1**) [2]. Therefore, vitreous proteomics is a valuable and feasible tool for expanding current understanding of the molecular milieu driving retinal disease. Ultimately, this method may elucidate novel therapeutic targets, resulting in better vision preservation and improved quality of life for patients affected by retinal disease.


Guei-Sheung Liu and Jiang-Hui Wang (eds.), *Diabetic Retinopathy: Methods and Protocols*, Methods in Molecular Biology, vol. 2678, https://doi.org/10.1007/978-1-0716-3255-0_11,

The vitreous proteome has a high dynamic range similar to that of blood and other extracellular fluids. A handful of highly abundant proteins account for more than half of the protein mass, while functionally important signaling proteins are present with abundances >6 orders of magnitude below them [7]. Quantifying the low-abundance proteome is a critical but challenging goal in vitreous proteomics (*see* **Note 2**). In order for vitreous proteomics to achieve the sensitivity required to achieve its clinical goals, rigorous methodology must be employed. Here, we discuss the selection, storage, and processing of samples, mass spectrometry (MS), and data analysis.

## 2 Materials

### 2.1 Protein Concentration Assay

The Pierce BCA Protein Assay Kit is recommended for protein concentration analysis. The Pierce Quantitative Fluorometric Peptide Assay can be used for peptide quantification if necessary.

### 2.2 SDS-PAGE Analysis

Pre-cast Invitrogen NuPAGE™ 4 to 12%, Bis-Tris, 1.0–1.5 mM gels are recommended. Invitrogen SYPRO™ Ruby Protein Gel Stain can be used to stain the gels.

### 2.3 Sample Fractionation

The recommended system for pre-fractionation of peptide samples prior to LC-MS/MS is the Pierce™ High pH Reversed-Phase Peptide Fractionation Kit.

### 2.4 Digestion

High-quality trypsin is essential. Sequencing Grade Modified Trypsin from Promega is a common choice. Mass spectrometry-grade Lys-C from Promega can also be used. Dithiothreitol and iodoacetamide from Sigma are recommended. HPLC grade water should be used for digestion and all subsequent steps.

### 2.5 Abundant Protein Removal

Pierce High Select™ Top 14 Depletion kits or MARS abundant protein depletion kits are common choices for abundant protein depletion.

### 2.6 LC-MS/MS Reagents

HPLC grade water and acetonitrile are required for LC-MS/MS analysis. EMD Millipore formic acid of the highest purity available is recommended.

## 3 Methods

### 3.1 Sample Selection

Human vitreous can be obtained from postmortem donors, in clinic from living patients, or in the operating room as part of a pars plana vitrectomy [1, 2, 6, 8], and each source has advantages and disadvantages. The benefit of postmortem eyes is that there is no risk to the patient. However, it is possible that the vitreous proteome in donor eyes is influenced by retinal ischemia following death, and it is difficult to predict proteome changes affected by the time from death to organ harvesting. Donor vitreous has also been shown to contain nearly 30 times the number of EVs than that of living patients, though the reason for this and its effect on protein composition are unclear [2]. Clinical vitreous samples are an important option for the study of early-stage retinal disease or diseases without surgical indications. Multiple studies have shown that vitreous can be easily and safely obtained in clinic [6, 8]. The main disadvantages to this source include a small risk of retinal tear or detachment during vitreous aspiration and possible clogging of the biopsy device, preventing successful sample acquisition. Obtaining vitreous as part of pars plana vitrectomy is perhaps the most common method and has the benefit of repurposing surgical waste. An advantage of this method is that it requires minimal effort and does not confer any additional risk to the patient.

Regardless of the source of vitreous, care should be taken to avoid variables that may confound the vitreous proteome. We recommend documenting patients' age, presence or absence of a posterior vitreous detachment, and intraocular lens status, as these variables may affect the structure and/or protein content of vitreous [9, 10]. We also recommend controlling for current or prior major eye disease, such as glaucoma, DR, AMD, or uveitis, as these conditions are known to alter the vitreous proteome [11–17]. It is important to also consider prior procedures such as intravitreal injection or pan-retinal photocoagulation and the time from procedure to sample collection in certain populations, as this may aid in interpreting findings between patients. Investigators may also choose to control for major systemic diseases, such as diabetes or cancer, depending on the goal of a given study, though the degree to which such conditions impact the vitreous proteome is not entirely clear. Further, history of vitrectomy renders vitreous unusable for most studies, as vitreous gel does not regenerate following removal [18].

The ideal control group for proteomic studies of retinal disease would consist of age-matched controls with no history of major ocular or systemic disease. In practice, however, the most commonly used control group consists of vitreous from patients undergoing pars plana vitrectomy for macular hole (MH) and/or epiretinal membrane (ERM) repair. MH and ERM are anatomical abnormalities of the retina that usually occur as sequelae of a posterior vitreous detachment. Therefore, these conditions should mainly involve structural rather than molecular changes in the retina, making the effect on the vitreous proteome minimal.

### 3.2 Sample Size

Prior proteomic studies of vitreous have used a wide range of sample sizes, from as few as 1 sample per group to >70 per group [4, 19], presumably based on availability rather than statistical power. However, a recent study performed the first power analysis for mass spectrometry-based vitreous proteomics in proliferative diabetic retinopathy (PDR), demonstrating that for proteins with the highest effect sizes, a sample size of $n = 8–10$ is required to achieve a power of about 0.8 [5]. Further work is needed to determine whether these findings are generalizable to other retinal diseases.

### 3.3 Sample Storage

Sample storage is critically important in preserving sample integrity between acquisition and analysis. Immediately upon acquisition, vitreous should be placed on wet ice and transferred to a −80 °C freezer until ready to be used.

### 3.4 Sample Validation

Before subjecting vitreous samples to MS-based proteomics, several steps can aid in validating sample integrity. First, it is important to measure protein concentration. For investigators interested in proteins contained within vitreous EVs, vitreous should be incubated with a buffer that can lyse EVs, releasing membrane-bound intraluminal proteins. One such buffer is radioimmunoprecipitation assay (RIPA) buffer. If interested in the quantity of EVs, nanoparticle tracking analysis can be performed prior to EV lysis. Next, samples should be subjected to sodium dodecyl sulfate-polyacrylamide gel electrophoresis (SDS-PAGE) and stained for protein to confirm prominent bands at 55 (transthyretin) and 64 kDa (albumin), which are normal components of unfractionated vitreous. These and other plasma proteins are abundant in vitreous. Therefore, abundant protein depletion should be performed to remove abundant plasma proteins from vitreous, thus increasing the depth of proteome coverage. Protein concentration should again be measured following depletion in order to calculate protein recovery rate and depletion efficiency. Expected protein recovery rate is approximately 10–20%, with a depletion efficiency of 80–90%. SDS-PAGE should then be repeated to confirm the absence of the transthyretin and albumin bands from post-depletion vitreous; these bands should be present in one of the subsequent elution fractions. Following these steps, samples are ready for allocation into appropriate volumes for MS.

### 3.5 Sample Fractionation

Sample fractionation is critical to achieving high dynamic range and quantifying low-abundance vitreous proteins. It can be incorporated at various points of the analytical pipeline. Following abundant protein depletion, proteins can be fractionated by one- or two-dimensional gel electrophoresis (although this method is now rarely used due to the difficulty of extracting proteins/peptides from the gel) or by liquid chromatography. After digestion (described below), peptides can be subjected to one- or two-dimensional chromatography to separate them. For protein or peptide fractionation, splitting the sample into more fractions increases the proteome depth of coverage at the expense of increasing the amount of instrument time required for analysis.

### 3.6 Sample Digestion

Once fractionation is complete, proteins are digested. Trypsin is by far the most common enzyme used for digestion. Its dominance in proteomic sample preparation is due to the convergence of a few properties. It acts quickly and with minimal non-specific cleavages. Further, it cleaves proteins' C-terminal (on the right) of lysine and arginine residues which has two benefits: (1) those residues have appropriate frequency in the proteome to generate peptides that are optimally sized for mass spectrometry, and (2) they are positively charged amino acids. Most tryptic peptides at neutral or lower pH will therefore have one positive charge on their N-terminus (left end) and another on their C-terminus (right end). Two positive charges (2+) is the optimum number for peptide fragmentation in the mass spectrometer. Peptides with one or three charges are less likely to be identified. Another commonly used option is Lys-C [20] which cleaves C-terminal to lysine residues, similarly leading to peptides with N- and C-terminal positive charges.

### 3.7 Liquid Chromatography

Mass spectrometry is typically combined with reversed-phase liquid chromatography (LC) which is used to simplify the peptide mixture as much as possible before introduction into the mass spectrometer. Key properties of the LC system are flow rate and pressure. Low flow rates result in better sensitivity, which is important where material is limited but imposes a cost in increased maintenance and troubleshooting time. Typical low flow rates are near 1 μL/min. LC systems that can generate higher pressure can use longer columns and/or smaller particles relative to lower pressure systems resulting in increased separation efficiency. Modern LC systems generate 15,000 psi.

LC also helps to put the peptides into a MS-compatible solution. A typical LC buffer consists of acetonitrile (0–40%) and formic acid (0.1%) in water. Volatility and acidic pH are important properties of this solution. This brings up another benefit of the property of tryptic peptides mostly having a uniform charge state (2+): they have a uniform elution profile in reversed-phase LC.

### 3.8 Mass Spectrometry

MS combined with LC can identify and quantify proteins across a dynamic range of more than four orders of magnitude. The main components of a mass spectrometer are an ion source, a mass analyzer, and a detector. Together, these apparatuses function to separate ionized sample components in gas phase according to their mass-to-charge (m/z) ratio via electric fields, magnetic fields, or both. This yields a mass spectrum of signal intensity and m/z value [21], allowing the identification and quantification of peptides.

For MS, biological samples must be converted into the gas phase, and their analytes must be ionized. Electrospray ionization (ESI) is the most common ionization technique. It relies on a liquid solvent and is therefore usually coupled to LC systems as described above [22]. The input to the ESI process is peptides in an acidic and volatile solution, and the output is positively charged peptides in the gas phase.

Mass analyzers make up the main part of a mass spectrometer. The most common types of mass analyzers used in proteomics include time of flight (TOF), quadrupole, ion trap, and orbitrap [22, 23]. Key properties of the analyzers are duty cycle (speed) and resolution. High resolution means that peaks that are close to each other can still be distinguished, and higher resolution is preferable to lower resolution. Orbitrap analyzers can achieve high resolution but at a cost of long (~1 s) scan times. In contrast, ion trap analyzers are fast (scan time < 100 ms) but generate lower-resolution data. The reader should be aware that multiple types of mass analyzer can be combined to take advantage of their different benefits. Mass analyzers are associated with a detector, and the data from the detector are transformed into mass spectra.

Proteomics samples are typically analyzed using tandem mass spectrometry (MS/MS). Peptide abundance and m/z are measured (termed the MS1 scan), and then high-abundance peptides are isolated and fragmented, followed by abundance and m/z measurement of resulting fragments (MS2 scan). The combination of parent ion and fragment ion m/z is important for peptide identification. Typically, a single MS1 scan will be followed by multiple MS2 scans of the most abundant peptides determined in real time by the mass spectrometer. This process of choosing peptides for MS2 "on the fly" is termed data-dependent acquisition (DDA). DDA is a valuable tool for identifying low-abundance proteins' high dynamic range samples such as vitreous, but it means that the set of peptides identified by mass spectrometry for two samples, even if they are identical, won't be the same. This introduces a missing data problem that will be discussed below. The time it takes a mass spectrometer to perform an MS1 scan and associated MS2 scans is termed the duty cycle. A duty cycle <2 s is expected for modern mass spectrometers, and most can generate >20 MS2 scans per second. Note that the width of the chromatographic peaks eluting from a typical LC system will be >10 s, so multiple analysis cycles can be performed while a peptide elutes.

### 3.9 Quantification

In quantitative proteomics, the abundance of a given protein is measured and can be compared across samples or experimental groups. Different methods exist for measuring abundance, each with its own pros and cons. For example, sample prep for label-free quantification is quite simple and inexpensive. However, samples must be run individually, which increases cost and necessitates normalization across samples. In contrast, isobaric tagging entails more complicated sample prep but enables multiplexing many samples for a single run, thereby reducing cost [24, 25]. Multiplexing is a useful strategy in light of the sample sizes required for sufficient statistical power. Ten- and 16-plex kits are commonly used for this purpose.

### 3.10 Data Analysis

Mass spectrometry data analysis aims to use the MS/MS spectra to reconstitute the original protein composition of a sample. It involves multiple steps, including (1) assigning peptide sequences to mass spectra, (2) statistical scoring of peptide-spectrum matches, and (3) identification of proteins from the filtered peptide list [24].

The most commonly used method for identifying peptides from mass spectra is database searching, which compares experimental spectra to theoretical spectra. The theoretical spectra are generated by in silico digestion of a protein sequence database using the same enzyme used for protein digestion. Popular protein sequence databases include RefSeq [26] and UniProt [27]. Based on how closely the experimental spectra resemble the theoretical spectra, search scores are assigned to spectral matches. Search parameters can be defined to aid the delineation of true vs. false matches [24]. Searches against theoretical spectra generated from a protein database where all protein sequences have been reversed (reversed database search) are used to define the range of scores typical for false positives.

Database searching yields multiple candidate peptides ranked by the search score, with the most likely match generally being the top scoring candidate. The validity of a single peptide-spectrum match (PSM) cannot be conclusively determined without further experiments. However, the global fraction of invalid PSMs in a proteomics dataset, termed the false discovery rate (FDR), can be calculated using the reversed database search results. Candidate PSMs from the combined forward and reversed database search results are ranked. For each PSM score, the fraction of reversed database hits with a higher score is recorded and converted to an FDR. The researcher chooses the score associated with their desired FDR (typically 1%) as the threshold for valid peptide identifications [24].

Depending on the goal of the experiment, other metrics can be applied to assess MS performance. After a high-quality peptide list is generated, proteins can be inferred, and this step is usually built into the search database used to generate the peptide list. Protein inference is commonly done to generate the smallest possible protein list that can explain all identified peptides [28]. The protein list is then ready for data visualization and interpretation.

Missing values are an obstacle for quantitative proteomics data analysis. The structure of the missing values depends on how the data were generated. As noted above, data-dependent acquisition means that proteins have some probability of dropping out of an analysis even if they are truly present in a sample. Lower-abundance proteins have a higher chance of dropout. Multiplexed analyses have almost no dropout within a multiplex run, but dropout between runs remains a problem. It is important to use an analysis system that properly handles missing values, for example, Limma [29] which infers variance from similar proteins to allow statistical evaluation of proteins with only two quantitative values. Alternatively, one may choose to impute missing data so as not to lose values across experiments.

A valuable tool for downstream proteomic analysis is differential expression analysis, which can compare protein expression across experimental groups. A variety of statistical tests can be used to determine which differences in protein expression are statistically significant, and the specific test chosen will depend on the experiment. Researchers can repurpose statistical analysis packages designed for mRNA analysis [29] or packages specifically targeted to proteomics [30].

Rather than comparing differences in individual proteins, it is often helpful to place differential expression results into a functional context using gene set enrichment and/or pathway analyses. A common first step in data interpretation is Gene Ontology (GO) [http://geneontology.org] [31]. This tool assigns hierarchically clustered terms from three categories (biological processes, molecular functions, and cellular components) to the dataset. Enrichment analysis can show which GO terms are most strongly represented in a given sample or group [32]. This and other types of annotation are often done with DAVID (https://david.ncirf.gov). Another way to analyze data is in terms of protein interactions. This can be done with STRING (https://string-db.org), which includes built-in literature mining and GO classification. The types of interactions available in the STRING database include both known physical and/or functional interactions and predicted interactions. For enhanced visualization of protein-protein interactions, STRING data can easily be imported into Cytoscape [33], which allows more customized visualization. Data can also be analyzed by pathway analysis, which identifies upregulated and downregulated functional pathways based on the presence of pathway components and their actions (e.g., activating vs. inhibiting) on the overall pathway function. Common pathway databases include KEGG (https://www.kegg.jp), Reactome (https://reactome.org), Ingenuity Pathway Analysis (Qiagen), BioCarta (http://www.biocarta.com), PANTHER (http://www.pantherdb.org), and GenMAPP (http://www.genmapp.org) [34–38]. Ingenuity Pathway Analysis (Qiagen, Hilden, Germany) and iPathway Guide (Advaita, Ann Arbor, MI) are popular pathway analysis tools that incorporate GO annotation, interaction networks, and other analytic functions.

## 4 Notes

1. The resulting EV sample is a rich source of biological information [2] and has a dynamic range that is tractable for MS analysis [39, 40]. Alternatively, antibody-based approaches have recently shown promise in quantifying low-abundance proteins in biofluids [41].
2. Quantifying low-abundance proteins in biofluids has been a challenge that MS-based proteomics has been struggling with for decades.

### References


1. Gardner TW, Sundstrom JM (2017) A proposal for early and personalized treatment of diabetic retinopathy based on clinical pathophysiology and molecular phenotyping. Vis Res. https://doi.org/10.1016/j.visres.2017.03.006
2. Zhao Y, Weber SR, Lease J, Russo M, Siedlecki CA, Xu LC, Chen H, Wang W, Ford M, Simo R, Sundstrom JM (2018) Liquid biopsy of vitreous reveals an abundant vesicle population consistent with the size and morphology of exosomes. Transl Vis Sci Technol 7(3):6. https://doi.org/10.1167/tvst.7.3.6
3. Skeie JM, Roybal CN, Mahajan VB (2015) Proteomic insight into the molecular function of the vitreous. PLoS One 10(5):e0127567. https://doi.org/10.1371/journal.pone.0127567
4. Loukovaara S, Nurkkala H, Tamene F, Gucciardo E, Liu X, Repo P, Lehti K, Varjosalo M (2015) Quantitative proteomics analysis of vitreous humor from diabetic retinopathy patients. J Proteome Res 14(12):5131–5143. https://doi.org/10.1021/acs.jproteome.5b00900
5. Weber SR, Zhao Y, Ma J, Gates C, da Veiga Leprevost F, Basrur V, Nesvizhskii AI, Gardner TW, Sundstrom JM (2021) A validated analysis pipeline for mass spectrometry-based vitreous proteomics: new insights into proliferative diabetic retinopathy. Clin Proteomics 18(1):28. https://doi.org/10.1186/s12014-021-09328-8
6. Ghodasra DH, Fante R, Gardner TW, Langue M, Niziol LM, Besirli C, Cohen SR, Dedania VS, Demirci H, Jain N, Jayasundera KT, Johnson MW, Kalyani PS, Rao RC, Zacks DN, Sundstrom JM (2016) Safety and feasibility of quantitative multiplexed cytokine analysis from office-based vitreous aspiration. Invest Ophthalmol Vis Sci 57(7):3017–3023. https://doi.org/10.1167/iovs.15-18721
7. Anderson NL, Anderson NG (2002) The human plasma proteome: history, character, and diagnostic prospects. Mol Cell Proteomics 1(11):845–867. https://doi.org/10.1074/mcp.r200007-mcp200
8. Pfahler SM, Brandford AN, Glaser BM (2009) A prospective study of in-office diagnostic vitreous sampling in patients with vitreoretinal pathology. Retina 29:1032–1035. https://doi.org/10.1097/IAE.0b013e3181a2c1eb
9. Sebag J (1987) Ageing of the vitreous. Eye 1 (Pt 2):254–262. https://doi.org/10.1038/eye.1987.45
10. Sebag J (1989) Development and aging of the vitreous. In: The vitreous: structure, function, and pathobiology. Springer, New York, pp 73–95

11. Mirzaei M, Gupta VB, Chick JM, Greco TM, Wu Y, Chitranshi N, Wall RV, Hone E, Deng L, Dheer Y, Abbasi M, Rezaeian M, Braidy N, You Y, Salekdeh GH, Haynes PA, Molloy MP, Martins R, Cristea IM, Gygi SP, Graham SL, Gupta VK (2017) Age-related neurodegenerative disease associated pathways identified in retinal and vitreous proteome from human glaucoma eyes. Sci Rep 7(1): 12685. https://doi.org/10.1038/s41598-017-12858-7
12. Angi M, Kalirai H, Coupland SE, Damato BE, Semeraro F, Romano MR (2012) Proteomic analyses of the vitreous humour. Mediat Inflamm 2012:148039. https://doi.org/10.1155/2012/148039
13. Walia S, Clermont AC, Gao B-B, Aiello LP, Feener EP (2010) Vitreous proteomics and diabetic retinopathy. Semin Ophthalmol 25: 289–294. https://doi.org/10.3109/08820538.2010.518912
14. Weber SR, Zhao Y, Gates C, Ma J, da Veiga Leprevost F, Basrur V, Nesvizhskii AI, Gardner TW, Sundstrom JM (2021) Proteomic analyses of vitreous in proliferative diabetic retinopathy: prior studies and future outlook. J Clin Med 10(11). https://doi.org/10.3390/jcm10112309
15. Nobl M, Reich M, Dacheva I, Siwy J, Mullen W, Schanstra JP, Choi CY, Kopitz J, Kretz FT, Auffarth GU, Koch F, Koss MJ (2016) Proteomics of vitreous in neovascular age-related macular degeneration. Exp Eye Res 146:107–117. https://doi.org/10.1016/j.exer.2016.01.001
16. Pfahler SM, Brandford AN, Glaser BM (2009) A prospective study of in-office diagnostic vitreous sampling in patients with vitreoretinal pathology. Retina 29(7):1032–1035
17. Kasudhan KS, Sarkar S, Gupta V, Gupta A, Chakraborti A (2018) Identification of unique proteins in vitreous fluid of patients with non-infectious uveitis. Acta Ophthalmol 96(8): e989–e1003. https://doi.org/10.1111/aos.13801
18. Bishop PN (2000) Structural macromolecules and supramolecular organisation of the vitreous gel. Prog Retin Eye Res 19(3):323–344
19. Koyama R, Nakanishi T, Ikeda T, Shimizu A (2003) Catalogue of soluble proteins in human vitreous humor by one-dimensional sodium dodecyl sulfate-polyacrylamide gel electrophoresis and electrospray ionization mass spectrometry including seven angiogenesis-regulating factors. J Chromatogr B Anal Technol Biomed Life Sci 792(1):5–21
20. Glatter T, Ludwig C, Ahrne E, Aebersold R, Heck AJ, Schmidt A (2012) Large-scale quantitative assessment of different in-solution protein digestion protocols reveals superior cleavage efficiency of tandem Lys-C/trypsin proteolysis over trypsin digestion. J Proteome Res 11(11):5145–5156. https://doi.org/10.1021/pr300273g
21. Gross JH (2017) Mass spectrometry: a textbook, 3rd edn. Springer, Cham
22. Aebersold R, Mann M (2003) Mass spectrometry-based proteomics. Nature 422(6928):198–207
23. Michalski A, Damoc E, Hauschild JP, Lange O, Wieghaus A, Makarov A, Nagaraj N, Cox J, Mann M, Horning S (2011) Mass spectrometry-based proteomics using Q exactive, a high-performance benchtop quadrupole Orbitrap mass spectrometer. Mol Cell Proteomics 10(9):M111.011015. https://doi.org/10.1074/mcp.M111.011015
24. Chen C, Hou J, Tanner JJ, Cheng J (2020) Bioinformatics methods for mass spectrometry-based proteomics data analysis. Int J Mol Sci 21(8). https://doi.org/10.3390/ijms21082873
25. Bantscheff M, Schirle M, Sweetman G, Rick J, Kuster B (2007) Quantitative mass spectrometry in proteomics: a critical review. Anal Bioanal Chem 389(4):1017–1031. https://doi.org/10.1007/s00216-007-1486-6
26. Pruitt KD, Tatusova T, Maglott DR (2007) NCBI reference sequences (RefSeq): a curated non-redundant sequence database of genomes, transcripts and proteins. Nucleic Acids Res 35 (Database issue):D61–D65. https://doi.org/10.1093/nar/gkl842
27. UniProt C (2019) UniProt: a worldwide hub of protein knowledge. Nucleic Acids Res 47 (D1):D506–D515. https://doi.org/10.1093/nar/gky1049
28. Nesvizhskii AI, Keller A, Kolker E, Aebersold R (2003) A statistical model for identifying proteins by tandem mass spectrometry. Anal Chem 75(17):4646–4658. https://doi.org/10.1021/ac0341261
29. Ritchie ME, Phipson B, Wu D, Hu Y, Law CW, Shi W, Smyth GK (2015) Limma powers differential expression analyses for RNA-sequencing and microarray studies. Nucleic Acids Res 43(7):e47. https://doi.org/10.1093/nar/gkv007
30. Choi M, Chang CY, Clough T, Broudy D, Killeen T, MacLean B, Vitek O (2014) MSstats: an R package for statistical analysis of quantitative mass spectrometry-based

proteomic experiments. Bioinformatics 30(17):2524–2526. https://doi.org/10.1093/bioinformatics/btu305

31. Ashburner M, Ball CA, Blake JA, Botstein D, Butler H, Cherry JM, Davis AP, Dolinski K, Dwight SS, Eppig JT, Harris MA, Hill DP, Issel-Tarver L, Kasarskis A, Lewis S, Matese JC, Richardson JE, Ringwald M, Rubin GM, Sherlock G (2000) Gene ontology: tool for the unification of biology. The gene ontology consortium. Nat Genet 25(1):25–29. https://doi.org/10.1038/75556
32. Schmidt A, Forne I, Imhof A (2014) Bioinformatic analysis of proteomics data. BMC Syst Biol 8(Suppl 2):S3. https://doi.org/10.1186/1752-0509-8-S2-S3
33. Szklarczyk D, Morris JH, Cook H, Kuhn M, Wyder S, Simonovic M, Santos A, Doncheva NT, Roth A, Bork P, Jensen LJ, von Mering C (2017) The STRING database in 2017: quality-controlled protein-protein association networks, made broadly accessible. Nucleic Acids Res 45(D1):D362–D368. https://doi.org/10.1093/nar/gkw937
34. Croft D, O'Kelly G, Wu G, Haw R, Gillespie M, Matthews L, Caudy M, Garapati P, Gopinath G, Jassal B, Jupe S, Kalatskaya I, Mahajan S, May B, Ndegwa N, Schmidt E, Shamovsky V, Yung C, Birney E, Hermjakob H, D'Eustachio P, Stein L (2011) Reactome: a database of reactions, pathways and biological processes. Nucleic Acids Res 39 (Database issue):D691–D697. https://doi.org/10.1093/nar/gkq1018
35. Kanehisa M, Goto S (2000) KEGG: Kyoto encyclopedia of genes and genomes. Nucleic Acids Res 28(1):27–30. https://doi.org/10.1093/nar/28.1.27
36. Kanehisa M, Goto S, Sato Y, Furumichi M, Tanabe M (2012) KEGG for integration and interpretation of large-scale molecular data sets. Nucleic Acids Res 40(Database issue): D109–D114. https://doi.org/10.1093/nar/gkr988
37. Mi H, Guo N, Kejariwal A, Thomas PD (2007) PANTHER version 6: protein sequence and function evolution data with expanded representation of biological pathways. Nucleic Acids Res 35(Database issue):D247–D252. https://doi.org/10.1093/nar/gkl869
38. Salomonis N, Hanspers K, Zambon AC, Vranizan K, Lawlor SC, Dahlquist KD, Doniger SW, Stuart J, Conklin BR, Pico AR (2007) GenMAPP 2: new features and resources for pathway analysis. BMC Bioinformatics 8:217. https://doi.org/10.1186/1471-2105-8-217
39. Burton JB, Carruthers NJ, Stemmer PM (2021) Enriching extracellular vesicles for mass spectrometry. Mass Spectrom Rev. https://doi.org/10.1002/mas.21738
40. Burton JB, Carruthers NJ, Hou Z, Matherly LH, Stemmer PM (2022) Pattern analysis of organellar maps for interpretation of proteomic data. Proteomes 10(2). https://doi.org/10.3390/proteomes10020018
41. Petrera A, von Toerne C, Behler J, Huth C, Thorand B, Hilgendorff A, Hauck SM (2021) Multiplatform approach for plasma proteomics: complementarity of Olink proximity extension assay technology to mass spectrometry-based protein profiling. J Proteome Res 20(1):751–762. https://doi.org/10.1021/acs.jproteome.0c00641

# Chapter 12

# Methods for Ex Vivo AAV Transduction in Rodent and Human Retinal Explants

Satheesh Kumar, Jiang-Hui Wang, and Guei-Sheung Liu

## Abstract

Translational research is heavily dependent on animal models, and reliable disease models are essential for the development of novel therapies. Here, we outline the methods for culturing mouse and human retinal explants. In addition, we show efficient adeno-associated virus (AAV) transduction of the mouse retinal explants to aid the study and development of AAV-based therapeutics against ocular diseases.

**Key words** Retinal explants, Adeno-associated virus

## 1 Introduction

Ex vivo models have been reported to closely resemble in vivo conditions as they retain neuronal connections and other non-neuronal cells [1]. They are also time- and cost-effective, offering an accessible disease model for studying retinal pathologies and performing preclinical studies for ocular therapies. While in vivo models are crucial for novel therapeutics to progress to clinical trials, their use is increasingly discouraged [2]. The development of retinal explants can therefore greatly facilitate translational work as it closely resembles conditions [1]. The availability of human retinal explants is also ideal for investigating novel therapeutics before initiation of clinical trial as rodent models do not accurately resemble the human retina and non-human primate models are costly and challenging to work with [2].

In addition, as adeno-associated viruses (AAV) are the preferred vectors for gene therapy development, their transduction in retinal explants would inform their capacity for gene delivery in humans. Here, we cultured mouse and human retinas and transduced the mouse retinal explant with AAV2/2[MAX] vectors [3] encoding a mCherry reporter. High transduction of the mouse retinal explant was observed a week after transduction, showing rapid onset of the

Guei-Sheung Liu and Jiang-Hui Wang (eds.), *Diabetic Retinopathy: Methods and Protocols*, Methods in Molecular Biology, vol. 2678, https://doi.org/10.1007/978-1-0716-3255-0_12,

AAV2/2[MAX] vectors. This model can be used to study the transduction capabilities of different AAV serotypes or variants in the retina, as well as obtain preclinical data for AAV-based ocular therapeutics.

In this chapter, we outline the production and purification of AAV constructs and their transduction in mouse retinal explants. This method will allow for highly reproducible results from animal and human tissues, offering an accessible model for translational studies.

## 2 Materials

### 2.1 AAV Production and Purification

1. HEK293D cells.
2. TE, pH 8.0, RNase-free.
3. 2 M $CaCl_2$.
4. HEPES-buffered saline, pH 7.0 (2× for transfection).
5. 0.15 M Na2HPO4.
6. TryPLE™ Express Enzyme (1×), phenol red.
7. Dulbecco's Modified Eagle Medium (DMEM), high glucose.
8. Iscove's Modified Dulbecco's Medium (IMDM).
9. Resuspension buffer.
   (a) NaCl (58.44 MW) 2.922 g (100 mM final).
   (b) MgCl2.6H2O (203.3 MW) 0.2033 g (2 mM final).
   (c) 1 M Tris-HCl (121.14 MW) 5 mL (10 mM final).
   (d) Make up to 500 mL Miili-Q water. Adjust to pH 8.0 and autoclave.
10. Corning® 500 $cm^2$ Square TC-Treated Culture Dish.
11. Corning® 100 mm TC-Treated Culture Dish.
12. Falcon® 14 mL Round Bottom High-Clarity PP Test Tube.
13. QIAquick Gel Extraction Kit (250).
14. Corning® Cell Lifter.
15. AAVpro Purification Kit.

### 2.2 Retinal Explant

1. Gibco™ Neurobasal™-A Medium, no D-glucose, no sodium pyruvate.
2. Gibco™ Penicillin-Streptomycin (100×).
3. Gibco™ L-Glutamine (200 mM).
4. Gibco™ B-27™ Supplement (50×), serum-free.
5. Gibco™ N-2 Supplement (100×).
6. Nunc™ Polycarbonate Cell Culture Inserts in Multi-well Plates (0.4 μm pore size).

#### 2.3 Cryosectioning and Immuno-cytochemistry

1. 3 mL disposable sterile Pasteur pipettes.
2. 4% paraformaldehyde in PBS.
3. 30% sucrose solution in PBS.
4. Optimal cutting temperature (OCT) medium.
5. Vinyl specimen mold trays, 15 mm × 15 mm × 5 mm (Tissue-Tek Cryomold).
6. Dry ice.
7. Cryotome.
8. Poly-L-lysine-coated glass slides.
9. 0.01 M phosphate-buffered saline (PBS): 0.95 g/L 10 mM phosphates as sodium phosphates, 0.201 g/L 2.68 mM KCL, and 8.12 g/L 140 mM NaCl, pH 7.45.
10. PBS/serum blocking solution: PBS supplemented with 0.1% Triton X-100 and 10% donkey serum.
11. PBS/Triton primary antibody solution: PBS supplemented with 0.1% Triton X-100, 2% donkey serum, and primary antibody.
12. PBS/Tween washing solution: PBS supplemented with 0.05% Tween 20.
13. PBS/Triton secondary antibody solution: PBS supplemented with 0.1% Triton X-100, 2% serum, and secondary antibody.
14. 12.3 mg/mL Hoechst 33342.
15. Antifade mounting reagent.

#### 2.4 Equipment

1. Tissue culture equipment (e.g., cell incubators, cell counter, water baths, and microscopes).
2. Centrifuges (e.g., high-speed centrifuge and microcentrifuge).
3. Quantitative polymerase chain reaction (qPCR) instrument (StepOnePlus™ Real-Time PCR System, Thermo Fisher Scientific).

## 3 Methods

#### 3.1 Production and Purification of rAAV

1. Seed HEK293D cells in 10 cm dishes at $7 \times 10^6$ cells per dish.
2. Ensure 80% confluency next day prior to transfection.
3. Two hours before transfection, remove old media, leaving 1 mL of media covering cell monolayer.
4. Add 9 mL of 2% IMDM +10% fetal bovine serum (FBS) media.

5. Transfect 5 μg vector, 5 μg packaging/capsid plasmid, and 15 μg of helper plasmid using calcium phosphate.
6. Incubate overnight at 37 °C.
7. Next day, replace old media with DMEM +2% FBS.
8. After 48–60 h transfection, harvest cells using cell scraper, and purify using Takara AAV2 Purification Kit.

### 3.2 Human Retinal Explant Culture

A donor human eyeball was dissected on ice within an hour of exenteration, as previously described [4].

1. Dissect the eyes to remove the iris, the lens, and the vitreous, and cut the sclera into four quadrants (*see* **Note 1**) (Fig. 1a).
2. Using a biopsy punch, trephine 4-mm-diameter discs of the neuroretina and retinal pigment epithelium (RPE)/choroid from the surrounding tissues (Fig. 1b).
3. Transfer both retinal and RPE/choroid explants to the inserts of a 24-well transwell plates (0.4 μm pore size) with culture medium of Neurobasal-A with phenol red supplemented with 1% penicillin/streptomycin, 0.8 mM L-glutamine, 2% B27 supplement, and 1% N2 supplement (Table 1; Fig. 1c).
4. Add 400 μL medium in the insert and 900 μL in well (*see* **Note 2**).

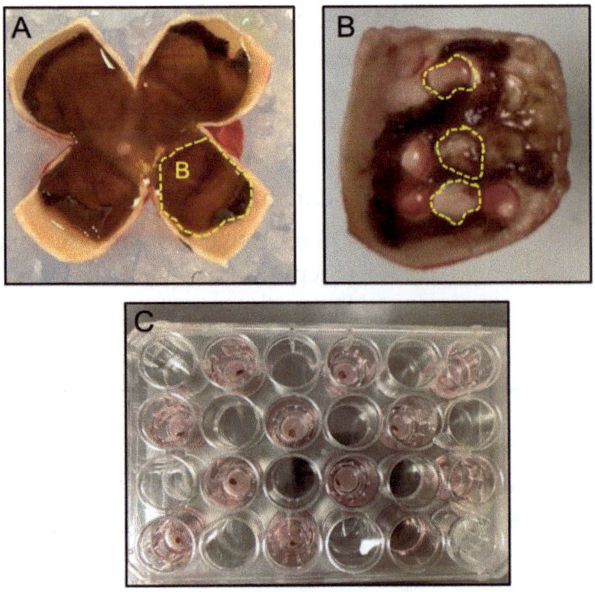


**Fig. 1** (**a**) Human sclera cut into quadrants. (**b**) Neuroretina and retinal pigment epithelium (RPE)/choroid trephined using a biopsy punch (indicate by yellow line). (**c**) Retinal explants cultured in inserts of 24-well plates

**Table 1**
**The formulation of retinal explant culture medium**

| Reagent | Amount (mL) |
|---|---|
| Neurobasal-A with phenol red | 96 |
| 100× Penicillin-streptomycin | 1 |
| 200 mM L-glutamine | 0.4 |
| 50× B27 supplement | 2 |
| 100× N2 supplement | 1 |

5. Culture neural retina and RPE/choroid explants in the same well for 24 h at 37 °C in 5% $CO_2$ incubator before AAV transduction. Change medium every 2 days.
6. Individually process neuroretina and RPE/choroid explants for immunostaining.

### 3.3 Mouse Retinal Explant Culture

1. Prepare retinal explant medium (Table 1).
2. Cull mice, enucleate the eyeballs using surgical forceps, place in Eppendorf tubes, and keep in ice.
3. Extract mice retina, split in half, and transfer each half-retina to individual culture inserts.
4. Add 400 μL of Neurobasal-A medium into each well.
5. Place individual culture inserts into 24-well plate (place tilted to avoid air trapping).
6. Add 300uL of Neurobasal-A medium into culture inserts to completely cover retinal explant.
7. Incubate plate overnight at 34 °C with 5% $CO_2$.

### 3.4 AAV Transduction

1. The day after culturing explant, add $1 \times 10^{10}$ vector genomes of AAV directly to culture inserts.
2. Return plate to incubator at 34 °C with 5% $CO_2$.
3. Every 48 h, aspirate old medium and add fresh medium as previously described (*see* **Note 3**).
4. At 6 days post-infection, check for retinal transduction using fluorescence microscope (Fig. 2).
5. Seven days post-infection, collect retinal explants using pipette in microcentrifuge tubes.
6. Centrifuge samples at maximum speed (≥14,000 rpm) for 2 min.
7. Discard supernatant and resuspend explant in RLT (RNA lysis) buffer.

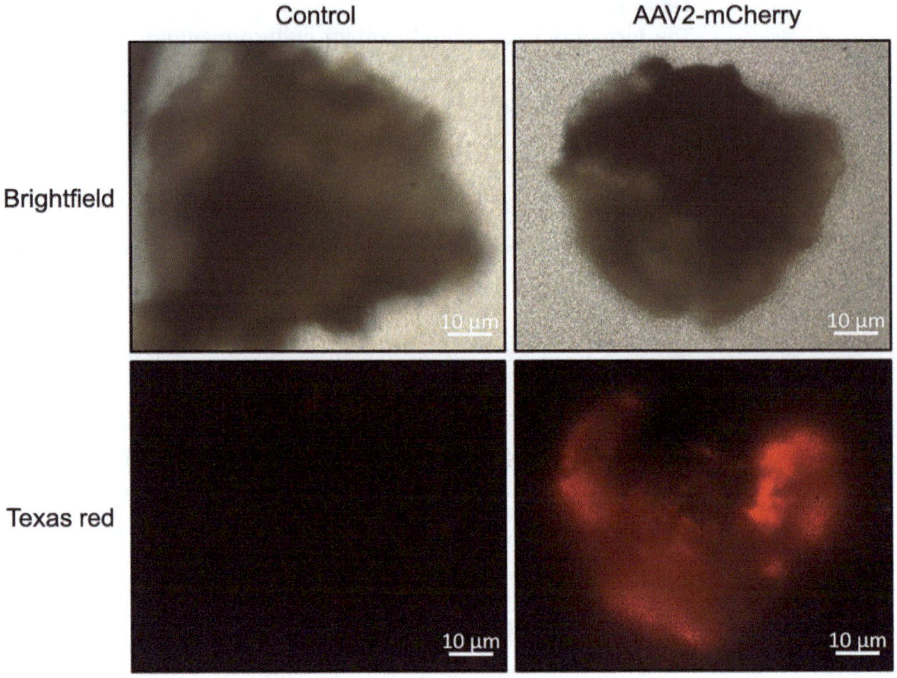


**Fig. 2** Representative fluorescence microscopy images of mouse retinal explant, 6 days post-AAV2-mCherry infection with Brightfield (top) and Texas Red (EX 560 nm, EM 635 nm; bottom) filters

8. Sonicate samples for 1–2 s to ensure the explant has dissolved.
9. Proceed with RNA extraction, cDNA synthesis, and qPCR analysis as previously described.

### *3.5 Cryosectioning and Immunocytochemistry*

1. Fix retinal explants in 4% paraformaldehyde in PBS, pH 7.4, for 1 h.
2. Transfer to 30% sucrose solution for at least 1 h for cryoprotection.
3. Transfer each fragment into PBS to briefly wash before transferring to individual vinyl specimen mold trays.
4. Remove excess PBS with pipette.
5. Fill tray with optimal cutting temperature medium ensuring each fragment is completely submerged and suitably orientated for sectioning.
6. Place on dry ice to freeze. Mark position of tissue and store at −80 °C before sectioning.
7. Section on cryotome to thickness of 12–18 μm onto poly-L-lysine-coated slides. Stores slides at −20 °C.
8. Wash slides with PBS 3 × 5min.
9. Block retinal sections for 1 h at room temperature in PBS/blocking solution.

10. Repeat washing step.
11. Incubate slides at 4 °C overnight in PBS/Triton primary antibody solution.
12. Wash for 3 × 5 min with PBS/Tween washing solution.
13. Incubate for 2 h at room temperature with species-appropriate PBS/Triton secondary antibody solution.
14. Counterstain sections with Hoechst 33342 1:5000 and mount with antifade reagent.
15. View on confocal with laser excitation wavelengths appropriate to the fluorophores used.

## 4 Notes

1. Use sterile equipment when harvesting retinas for culturing.
2. Media should be gently added to the culture inserts to ensure the explant is not disturbed.
3. Ensure proper aseptic techniques when working with viruses. Use 1% bleach to disinfect consumables before disposing.

### References


1. Murali A, Ramlogan-Steel CA, Andrzejewski S, Steel JC, Layton CJ (2019) Retinal explant culture: a platform to investigate human neuro-retina. Clin Exp Ophthalmol 47(2):274–285. https://doi.org/10.1111/ceo.13434. PMID: 30378239
2. Schnichels S, Paquet-Durand F, Löscher M, Tsai T, Hurst J, Joachim SC, Klettner A (2021) Retina in a dish: cell cultures, retinal explants and animal models for common diseases of the retina. Prog Retin Eye Res 81:100880. https://doi.org/10.1016/j.preteyeres.2020.100880
3. Reid CA, Ertel KJ, Lipinski DM (2017) Improvement of photoreceptor targeting via intravitreal delivery in mouse and human retina using combinatory rAAV2 capsid mutant vectors. Invest Ophthalmol Vis Sci 58(14): 6429–6439. https://doi.org/10.1167/iovs.17-22281
4. Wiley LA, Burnight ER, Kaalberg EE, Jiao C, Riker MJ, Halder JA, Luse MA, Han IC, Russell SR, Sohn EH, Stone EM, Tucker BA, Mullins RF (2017) Assessment of adeno-associated virus serotype tropism in human retinal explants. Hum Gene Ther 29(4):424–436. https://doi.org/10.1089/hum.2017.179

# Chapter 13

# Methods for RPE Sheet Construction Using iPS-CM in Conjunction with Natural Scaffold of Corneal Lenticule Derived from SMILE

**Jianing Gu, Yini Wang, Chengcheng Ding, Shibo Tang, and Jiansu Chen**

## Abstract

In vitro generation of a functional retinal pigment epithelium (RPE) monolayer sheet is useful and promising for RPE cell therapy. Here, we outline a method to construct engineered RPE sheets treated by induced pluripotent stem cell-conditioned medium (iPS-CM) in conjunction with femtosecond laser intrastromal lenticule (FLI-lenticule) scaffold to aid in enhanced RPE characteristics and cilium assembly. Such a strategy to construct RPE sheets is a promising avenue for developing RPE cell therapy, disease models, and drug screening tools.

**Key words** Retinal pigment epithelial cells, Tissue engineering, iPS-CM, Corneal lenticule

## 1 Introduction

Retinal pigment epithelium (RPE) cell therapy is one of the research directions for the treatment of RPE cell-related diseases in recent years, mainly including cell suspension injection and cell sheet transplantation [1, 2]. Although the approach of direct injection of RPE cell suspension into the posterior of the eye is relatively easy and has been used in clinical trials [1], an increasing number of reports show that transplantation of an engineered polarized RPE monolayer sheet promotes the survival and function of transplanted cells [3]. But how to enhance RPE characteristics after long-term culture in vitro and selecting a suitable scaffold for RPE sheet for better integration in vivo for RPE sheet construction are major challenges. Induced pluripotent stem cell-conditioned medium (iPS-CM) contains a substantial amount of cytokines, which promote cell proliferation and inhibit apoptosis, and prevent transformed morphology of RPE cells after passages, leading to better barrier function [4–6]. In addition, suitable RPE cell scaffolds can improve cell survival and stability and maintain cell polarity in vitro

Guei-Sheung Liu and Jiang-Hui Wang (eds.), *Diabetic Retinopathy: Methods and Protocols*, Methods in Molecular Biology, vol. 2678, https://doi.org/10.1007/978-1-0716-3255-0_13,

[3]. RPE cell therapy needs a suitable scaffold serving as a Bruch's membrane analog as Bruch's membrane is often disrupted in some retinal diseases [7, 8]. Here, we use the femtosecond laser intrastromal lenticule (FLI-lenticule) which is extracted during the corneal small incision lenticule extraction (SMILE) procedure for myopic correction as the RPE carrier scaffold. FLI-lenticule shows the biological and structural integrity of extracellular matrix architecture, which promotes characteristic mature morphology of RPE. Moreover, FLI-lenticule shows favorable subretinal biocompatibility to avoid the occurrence of rejection. iPS-CM in conjunction with FLI-lenticule scaffold can help constructing a tissue-engineered RPE sheet that exhibits the morphology and physiology similar to native RPE tissue [9].

In this chapter, we describe the protocol of isolation and culture of adult RPE cells, preparation of iPS-conditioned medium, and decellularization of FLI-lenticule. We also outline the steps required for the construction of a tissue-engineered RPE sheet with iPS-CM treatment and FLI-lenticule scaffold. Such engineered RPE sheets have the potential to be used for RPE transplantation, disease models, and drug screening.

## 2 Materials

### 2.1 Adult RPE Cell Preparation

1. Adult donor eye.
2. Dissecting microscope.
3. Ophthalmic tweezers and scissors.
4. 0.25% EDTA-trypsinase.
5. RPE complete medium: DMEM/F12 containing 10% fetal bovine serum and 1% penicillin-streptomycin (P/S).

### 2.2 The Culture of Urine-Induced Pluripotent Stem Cells (iPS Cells)

1. 1% (v/v) Matrigel.
2. Essential 8 (E8) medium.
3. 0.22 μm cell filter.

### 2.3 Decellularization

1. Femtosecond laser intrastromal lenticule (FLI-lenticule): lenticules from the corneal small incision lenticule extraction procedure for myopic correction.
2. Hypertonic saline: 1.5 M sodium chloride in $dH_2O$.
3. DNAse: 5 U/mL DNAse in PBS/normal saline.
4. RNAse: 50 mg/mL RNAse in PBS/normal saline.

## 3 Methods

### 3.1 Isolation and Culture of Adult RPE Cells

1. Obtain donor eyes for corneal transplant, and transport to the laboratory in sterile medium, within 24 h after death.
2. Under a dissecting microscope, remove fractional choroid, and detach the retina without RPE layer mechanically from the donated tissue.
3. Digest the remainder of donated tissue for 40–60 min with 0.25% EDTA-trypsinase at 37 °C (*see* **Note 1**).
4. Dilute with complete medium. Blow RPE layer gently and repeatedly. Cell suspensions were collected.
5. Collect cell suspension to a 15 mL conical tube and centrifugation for 5 min at 100×*g*. Resuspend RPE cells in complete medium.
6. Seed RPE cells into cell culture-treated plastics. Incubate at 37 °C in an incubator with 5% $CO_2$ and 95% humidity until 80% confluence was reached on day 5.
7. From passage 1 (P1), passage RPE cells using 0.25% EDTA-trypsinase, and seed at a density of $10^5$ cells/well in 24-well plates.

### 3.2 Passaging of iPS Cells

1. Matrigel-coated six-well plate. Thaw 500 μL of Matrigel on ice. Transfer the Matrigel to the 50 mL DPBS. Mix and add 1 mL of Matrigel per each well of a six-well plate or 10 μg/cm$^2$ with a cooled pipette. Leave the plates at 37 °C for 1 h. Seal the plates with Parafilm and store them at 4 °C for up to 2 weeks.
2. Prewarm Matrigel-coated plates, cell culture medium E8, and EDTA.
3. Aspirate and discard E8 culture.
4. Wash the iPS cells with DPBS.
5. Add 1 mL of EDTA and incubate it at 37 °C for 3 min (*see* **Note 2**).
6. Aspirate and discard the EDTA solution.
7. Add 1 mL of E8 medium, and gently blow no more than ten times with a 1 mL pipette tip. The size of the cell clumps should be 4–20 cells on average (*see* **Note 3**).
8. Prepare culture plates by aspirating the remaining Matrigel solution from the plates prepared in **step 2** and adding 2 mL of E8.
9. The cells passaged at a ratio of 1:10, and then put the plate in an incubator at 37 °C.

### 3.3 Preparation of iPS-Conditioned Medium

1. Culture induced pluripotent stem (iPS) cells in E8 medium at a 37 °C, 5% $CO_2$ incubator.
2. Derive supernatant of iPS cells from the E8 medium cultured with iPS cells for 24 h.
3. Filter supernatant with a 0.22 μm filter to remove dead cells and store at −80 °C.
4. iPS-conditioned medium (iPS-CM) was iPS supernatant mixed with RPE complete medium at the ratio of 1:2.
5. When cells confluence about 70–80%, use EDTA to passage.

### 3.4 Decellularization of FLI-Lenticule

1. Wash FLI-lenticule three times with PBS.
2. Incubate in hypertonic saline for 48 h with NaCl change after 24 h on a plant shaker (*see* **Note 4**).
3. Aspirate and discard the hypertonic saline.
4. Wash FLI-lenticule three times with PBS.
5. Treat with 5 U/mL DNAse and 50 mg/mL RNAse for 48 h [10].
6. Aspirate and discard the enzyme solution.
7. Wash lenticule with PBS for 72 h and change every 24 h.
8. Image the decellularized FLI-lenticule by hematoxylin and eosin (H&E) staining to verify that the cells are eliminated (Fig. 1a).
9. Image the decellularized FLI-lenticule by a scanning electron microscope (SEM) (Fig. 1b).

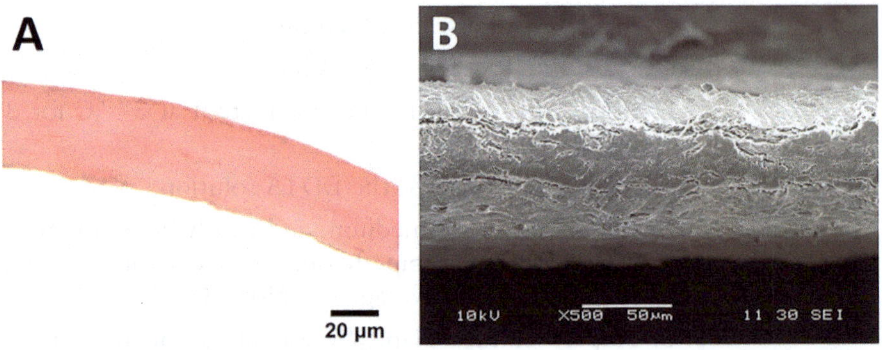


**Fig. 1** The structures of the decellularized FLI-lenticule scaffold. Decellularized FLI-lenticule maintains corneal ECM structure integrity without corneal stromal cells by H&E staining (**a**; scale bar: 20 μm); meanwhile, a cross-section displays the dense and crossed fibroid structure by SEM (**b**; scale bar: 50 μm). *FLI-lenticule* femtosecond laser intrastromal lenticule, *ECM* extracellular matrix, *SEM* scanning electron microscope

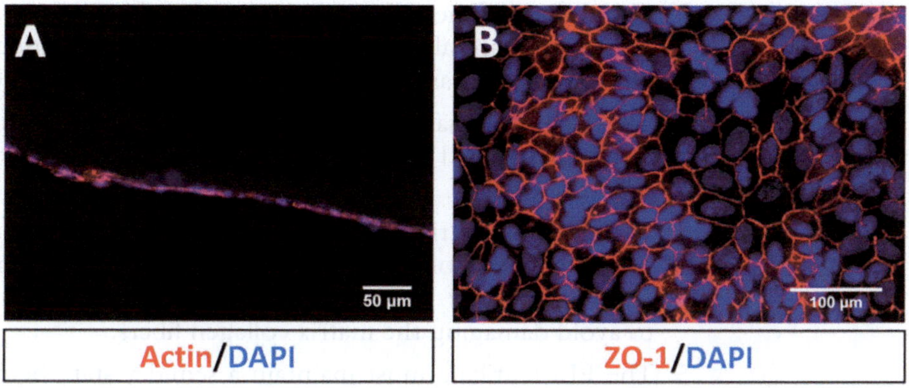


**Fig. 2** The structures of tissue-engineered RPE sheet. RPE cells are able to grow confluently and adhere firmly on FLI-lenticule by immunofluorescence staining images of actin (**a**; scale bar: 50 μm) and ZO-1 (**b**; scale bar: 100 μm). *RPE* retinal pigment epithelium

### 3.5 Construction of Tissue-Engineered RPE Sheet with iPS-CM Treatment and FLI-Lenticule Scaffold

1. Spread out acellular FLI-lenticule onto the 96-well culture plates, and place into the incubator or biological safety cabinetry until semidry (*see* **Note 5**).
2. Aspirate the supernatant of RPE cells and wash the cells three times with DPBS.
3. Aspirate PBS and add 0.5 mL of 0.25% EDTA-trypsinase per well of the six-well plate. Incubate the cells with trypsinase for 5–8 min at 37 °C.
4. Stop the reaction by adding 0.5 mL of RPE complete medium per well. Blow RPE layer gently and repeatedly. Cell suspensions were collected.
5. Collect cell suspension to a 15 mL conical tube and centrifugation for 5 min at 100×*g*. Resuspend RPE cells in a complete medium.
6. Seed the RPE cells ($10^3$ cells) treated with iPS-CM onto an acellular FLI-lenticule scaffold, and culture for 5 days.

4\. Harvest the tissue-engineered RPE sheet to make frozen sections for immunofluorescence staining of cross-section structure (Fig. 2a).

5\. Image the tissue-engineered RPE sheet by immunofluorescence staining of ZO-1 for their 3D architecture (Fig. 2b).

## 4 Notes

1. The digestion degree should take into consideration. After 20 min, we observe cells every 10 min and stop digestion when cells become round.

2. The timing of the enzymatic treatment varies, so it has to be monitored carefully. The EDTA should not be added for a prolonged time, as it might affect the viability of iPS cells.
3. This step is crucial for iPSC viability, as small cell clumps will result in increased death and big cell clumps impair differentiation efficiency.
4. The decellularization procedure was performed at room temperature under continuous agitation, and the plate shaker was set at a low speed. The rotational speed should not be too high to avoid damaging the matrix collagen fibers.
5. The FLI-lenticule must maintain a semidry state because the wet FLI-lenticule tends to float after seeding cells and the dry FLI-lenticule affects RPE cell viability.

## References


1. Seiler MJ, Aramant RB (2012) Cell replacement and visual restoration by retinal sheet transplants. Prog Retin Eye Res 31(6): 661–687. https://doi.org/10.1016/j.preteyeres.2012.06.003
2. Kashani AH, Lebkowski JS, Rahhal FM, Avery RL, Salehi-Had H, Chen S, Chan C, Palejwala N, Ingram A, Dang W, Lin CM, Mitra D, Pennington BO, Hinman C, Faynus MA, Bailey JK, Mohan S, Rao N, Johnson LV, Clegg DO, Hinton DR, Humayun MS (2021) One-year follow-up in a phase 1/2a clinical trial of an allogeneic RPE cell bioengineered implant for advanced dry age-related macular degeneration. Transl Vis Sci Technol 10(10): 13. https://doi.org/10.1167/tvst.10.10.13
3. Stern JH, Tian Y, Funderburgh J, Pellegrini G, Zhang K, Goldberg JL, Ali RR, Young M, Xie Y, Temple S (2018) Regenerating eye tissues to preserve and restore vision. Cell Stem Cell 23(3):453. https://doi.org/10.1016/j.stem.2018.08.014
4. Guo X, Chen Y, Hong T, Chen X, Duan Y, Li C, Ge R (2018) Induced pluripotent stem cell-derived conditional medium promotes Leydig cell anti-apoptosis and proliferation via autophagy and Wnt/beta-catenin pathway. J Cell Mol Med 22(7):3614–3626. https://doi.org/10.1111/jcmm.13641
5. Zhou Y, He Z, Gao Y, Zheng R, Zhang X, Zhao L, Tan M (2016) Induced pluripotent stem cells inhibit bleomycin-induced pulmonary fibrosis in mice through suppressing TGF-beta1/Smad-mediated epithelial to mesenchymal transition. Front Pharmacol 7:430. https://doi.org/10.3389/fphar.2016.00430
6. Lian RL, Guo XL, Chen JS, Guo YL, Zheng JF, Chen YW (2016) Effects of induced pluripotent stem cells-derived conditioned medium on the proliferation and anti-apoptosis of human adipose-derived stem cells. Mol Cell Biochem 413(1–2):69–85. https://doi.org/10.1007/s11010-015-2640-7
7. Warnke PH, Alamein M, Skabo S, Stephens S, Bourke R, Heiner P, Liu Q (2013) Primordium of an artificial Bruch's membrane made of nanofibers for engineering of retinal pigment epithelium cell monolayers. Acta Biomater 9(12):9414–9422. https://doi.org/10.1016/j.actbio.2013.07.029
8. Sugino IK, Rapista A, Sun Q, Wang J, Nunes CF, Cheewatrakoolpong N, Zarbin MA (2011) A method to enhance cell survival on Bruch's membrane in eyes affected by age and age-related macular degeneration. Invest Ophthalmol Vis Sci 52(13):9598–9609. https://doi.org/10.1167/iovs.11-8400
9. Gu J, Wang Y, Cui Z, Li H, Li S, Yang X, Yan X, Ding C, Tang S, Chen J (2019) The construction of retinal pigment epithelium sheets with enhanced characteristics and cilium assembly using iPS conditioned medium and small incision lenticule extraction derived lenticules. Acta Biomater 92:115–131. https://doi.org/10.1016/j.actbio.2019.05.017
10. Shafiq MA, Gemeinhart RA, Yue BY, Djalilian AR (2012) Decellularized human cornea for reconstructing the corneal epithelium and anterior stroma. Tissue Eng Part C Methods 18(5):340–348. https://doi.org/10.1089/ten.TEC.2011.0072

# Chapter 14

# Chromatin Immunoprecipitation Assay in Mouse Retinal Tissue

**Zhijie Wang, Shangli Ji, Jacey Hongjie Ma, Jiansu Chen, and Shibo Tang**

## Abstract

Chromatin immunoprecipitation (ChIP) is one of the most widely used methods for investigating interactions between proteins and DNA sequences. ChIP plays an important role in the transcriptional regulation study, which can locate the target genes of transcription factors and cofactors or monitor the sequence-specific genomic regions of histone modification. To analyze the interaction between transcription factors and several candidate genes, ChIP coupled with quantitative PCR (ChIP-PCR) assay is a basic tool. With the development of next-generation sequencing technology, ChIP-coupled sequencing (ChIP-seq) can provide the protein-DNA interaction information in a genome-wide dimension, which helps a lot in identifying new target genes. This chapter describes a protocol for performing ChIP-seq of transcription factors from retinal tissues.

**Key words** ChIP, ChIP-seq, Retina, Protein-DNA interaction

## 1 Introduction

Chromatin immunoprecipitation (ChIP) is a basic technology to investigate the interaction between a protein and a specific DNA sequence in live cells. For example, ChIP is widely used to identify the binding sites of transcription factors and map the genomic distribution of modified histone protein. A standard ChIP assay contains the following steps: treat live cells with the fixation reagent to crosslink protein-DNA, extract fragmented chromatin, immunoprecipitate the protein of interest which is combined with chromatin fragments with a specific antibody, reverse the crosslinking, and purify the DNA. The final product of the ChIP assay is a pool of fragmented DNA that is bound with the protein of interest. The ChIP-ed DNA then can be used for assessing the abundance of candidate DNA regions by quantitative PCR (ChIP-PCR). Alternatively, the precipitated DNA also can be subjected to a microarray chip (ChIP-chip) or next-generation sequencing (ChIP-seq),

Guei-Sheung Liu and Jiang-Hui Wang (eds.), *Diabetic Retinopathy: Methods and Protocols*, Methods in Molecular Biology, vol. 2678, https://doi.org/10.1007/978-1-0716-3255-0_14,

providing the genomic-wide binding pattern of chromatin-associated proteins. The enrichments of the fragmented DNA are referred to as "peak," which indicates the genomic region bound with the protein of interest at the moment of crosslinking.

With a high profile of sequencing, ChIP-seq is widely performed for whole retina tissue to reveal the gene regulatory networks of the transcription factors or histone modification in retinal development and retinal cell differentiation [1–3]. Because of the heterogeneity of retinal cells, some cell-specific epigenetic information might be lost in whole retina tissue, and some researchers performed ChIP-seq in specific cultured cells of the retina, such as retinal endothelial cells [4, 5]. Besides, ChIP-seq data can be combined with other omic data, such as transcriptome and chromatin accessibility data, to obtain a comprehensive profile of protein of interest or to identify direct transcriptional target genes [2, 6].

## 2 Materials

### 2.1 Retina Harvesting and Crosslinking

1. Carbon dioxide.
2. Fine scissors and forceps (autoclaved).
3. Pre-cold PBS (phosphate-buffered saline) buffer.
4. Light microscope.
5. Crosslinking buffer: 1% formaldehyde, 50 mM HEPES pH 7.5, 100 mM NaCl, 1 mM EDTA pH 8.0 and 0.5 mM EGTA pH 8.0, protease inhibitors.
6. 2.5 M glycine in ultrapure water.
7. 15 mL centrifuge tube.

### 2.2 Nuclei Isolation and Chromatin Sonication

1. Cell lysis buffer L1: 50 mM HEPES pH 7.5, 140 mM NaCl, 1 mM EDTA pH 8.0, 10% glycerol, 0.5% NP-40, 0.25% Triton X-100 and supplemented with protease inhibitors.
2. Wash buffer L2: 10 mM Tris-HCl pH 8.0, 200 mM NaCl, 1 mM EDTA pH 8.0, 0.5 mM EGTA pH 8.0 supplemented with protease inhibitors.
3. Nuclei lysis buffer L3: 10 mM Tris-HCl pH 8.0, 100 mM NaCl, 1 mM EDTA pH 8.0, 0.5 mM EGTA pH 8.0, 0.1% sodium deoxycholate and supplemented with protease inhibitors.
4. Sonicator.

#### 2.3 Chromatin Immunoprecipitation

1. Triton X-100.
2. Protein G magnetic beads.
3. 1.5 mL centrifuge tube (nuclease-free).
4. Block solution: 5 mg/mL BSA (bovine serum albumin) in ultrapure water.
5. Magnet rack.
6. ChIP-grade antibody of protein of interest.
7. Normal IgG control antibody.
8. RIPA ChIP wash buffer: 50 mM HEPES-KOH pH 8.0, 1 mM EDTA, 500 mM LiCl, 1% NP-40, 0.7% sodium deoxycholate.
9. TE buffer: 10 mM Tris-HCl pH 8.0, 1 mM EDTA pH 8.0.
10. ChIP elution buffer: 50 mM Tris-HCl pH 8.0, 10 mM EDTA pH 8.0, 1% SDS.

#### 2.4 Reverse Crosslinking and DNA Isolation

1. Proteinase K (20 mg/mL).
2. 5 M NaCl in water.
3. Phenol/chloroform, pH 8.0.
4. Chloroform/isoamyl alcohol (24:1).
5. Glycogen (20 mg/mL).
6. Ethanol.
7. Nuclease-free water.
8. Qubit 3.0 Fluorometer.

#### 2.5 Library Preparation and Sequencing

1. Commercial DNA library preparation kit (such as NEBNext® Ultra™ II DNA Library Prep Kit for Illumina®).
2. NovaSeq 6000 sequencer.

## 3 Methods

#### 3.1 Retina Harvesting and Crosslinking

1. Sacrifice mice by cervical dislocation or $CO_2$ asphyxiation. Pull out the eyes with forceps from killed mice immediately, and immerse the eye into cold PBS.
2. Under the microscope, cut the cornea with a fine scissor, and remove the iris, lens, and ciliary body. Hold the optic nerve or the sclera with forceps, and bring pressure on the sclera with other forceps to squeeze out the intact retina from the eye cup (*see* **Note 1**). Transfer retina tissues to a 15 mL centrifuge tube containing 1 mL 1% formaldehyde crosslinking buffer (*see* **Note 2**).

3. Pipette up and down heavily with a 1 mL tip to completely disaggregate the retina tissue (*see* **Note 3**). Add 4 mL of 1% formaldehyde crosslinking buffer, and incubate for 20 min at room temperature with rocking.
4. Add 275 μL of 2.5 M glycine to get a final concentration of 125 mM, and incubate for 5 min at room temperature with rocking to quench the crosslinking (*see* **Note 4**).
5. Centrifuge at 1000×*g* for 15 min at 4 °C. Discard the solution, and wash the cell pellet with 1 mL of pre-cold PBS containing protease inhibitors.
6. Centrifuge at 1000×*g* for 15 min at 4 °C and discard the supernatant. Repeat the wash three times. The cell pellets can be used for chromatin isolation directly or frozen at −80 °C until the further procedure.

### 3.2 Nuclei Isolation and Chromatin Sonication

1. Keep the cell pellet on ice and resuspend with 5 mL of cell lysis buffer L1. Rock at 4 °C for 10 min. Centrifuge the cells at 1000×*g* for 15 min at 4 °C.
2. Discard the supernatant and resuspend the cell pellet with 5 mL wash buffer L2. Rock at room temperature for 10 min. Centrifuge at 1000×*g* for 5 min at 4 °C. Discard the supernatant to get the nuclei pellet.
3. Resuspend nuclei pellet with 3 mL nuclei lysis buffer L3. Place the sample on ice for 10 min.
4. Keep the sample in an ice water bath. Sonicate the suspension of nuclei with a probe attached to the sonicator (*see* **Note 5**). Centrifuge at 15,000×*g* for 10 min at 4 °C to remove debris. Transfer the supernatant to a new 15 mL conical tube.
5. Take 50 μL lysate treated to reverse crosslinks, and purify the DNA to assess the overall size distribution of fragments (by 1% agarose gel electrophoresis) and the DNA concentration. Ideally, a smeared band can be observed, and 60–90% DNA fragments located between 250 bp and 1 kb.

### 3.3 Chromatin Immunoprecipitation

1. Add 333 μL of 10% Triton X-100 to the sonicated lysate. Mix by pipetting up and down. Save 50 μL of chromatin sample as input control, and store at −20 °C for further procedure.
2. Add protein G magnetic beads (100 μL for each ChIP reaction) to a 1.5 mL centrifuge tube. Add 1 mL of block solution (5 mg/mL BSA) to each tube to wash beads. Use a magnet rack to precipitate beads, aspirating the supernatant. Repeat washing the beads three times.
3. Add 1 mL of 5 mg/mL BSA solution to resuspend beads and add 10 μg of antibody or normal IgG control antibody and incubate overnight on a rotating platform at 4 °C.

4. Precipitate the beads with a magnetic rack and remove the supernatant. Add 1 mL of 5 mg/mL BSA solution to wash beads three times. After the last wash, add 100 μL of 5 mg/mL BSA solution to resuspend the beads.
5. Add antibody-bead complexes to the 1 mL of chromatin samples, and incubate overnight on a rotating platform at 4 °C.
6. Concentrate the beads with a magnetic rack and discard the supernatant.
7. Add 1 mL of pre-cold RIPA ChIP wash buffer to the tube, and wash the beads by inverting the tube. Repeat the wash seven times.
8. Add 1 mL of pre-cold TE buffer to wash beads once and discard the supernatant. Centrifuge the tubes for 3 min at 1000×*g* at 4 °C and remove the residual liquid.
9. Add 500 μL of ChIP elution buffer to each ChIP tube and resuspend the beads. Incubate at 65 °C for 15 min, and vortex the beads every 2 min. Centrifuge at 12,000×*g* for 1 min, and then transfer the supernatant to a new 1.5 mL centrifuge tube.

### 3.4 Reverse Crosslinking and DNA Isolation

1. Thaw the genomic input control on ice. Add 450 μL ChIP elution buffer to the input control tube. Incubate at 65 °C for 15 minutes. Add 5 μL of proteinase K and 11 μL of 5 M NaCl to each ChIP tube and the input tube. Vortex and incubate at 42 °C for 2 h. Then reverse crosslinks overnight at 65 °C incubations.
2. Add 500 μL of phenol/chloroform and vortex vigorously. Centrifuge at 15,000×*g* for 10 min at room temperature. Transfer the aqueous phase to a new tube. Add 500 μL of chloroform/isoamyl alcohol (24:1) and vortex. Centrifuge at 15,000×*g* for 10 min at room temperature (*see* **Note 6**).
3. Transfer the aqueous phase to a new 1.5 mL centrifuge tube. Add 1 μL glycogen, 1/10 volume of sodium acetate (3 M NaAc), and 2.5 volume of ethanol. Vortex and cool the tubes at −80 °C for at least 2 h.
4. Centrifuge at 15,000×*g* for 10 min at 4 °C and discard the supernatant. And wash the pellet with 1 mL of 70% ethanol, and centrifuge at 15,000×*g* for 10 min at 4 °C. Discard the supernatant and air-dry the pellet.
5. Resuspend the pellet with 20 μL of nuclease-free water. Quantify the concentration of DNA with fluorometric instruments (Qubit 3.0; *see* **Note 7**).

### *3.5 Library Preparation and Sequencing*

1. Prepare the sequencing library with commercial kits (such as SimpleChIP® ChIP-seq DNA Library Prep Kit, CST, or NEBNext® Ultra™ II DNA Library Prep Kit for Illumina®), following the manufacturer's instructions. The standard pipeline includes end repair of DNA fragments, dA-tailing, adaptor ligation, and library amplification (*see* **Note 8**). Then library products are finally sequenced on NovaSeq 6000 sequencer (Illumina) with paired-end mode.
2. After obtaining the raw sequence reads, quality control is required to filter low-quality reads and adaptors. The trimmed reads are then mapped to the reference genome. To discover the binding site of interest protein in the genome, peak calling is the most common approach. To gain insights of target protein, there are many downstream steps that can be performed. Firstly, peaks can be visualized in the Integrative Genomics Viewer (IGV) browser, which can demonstrate enriched regions of ChIP-seq signal in specific genomes. To further investigate the annotation of peaks, ChIPseeker R package can be applied to visualize the peaks' distribution and the binding pattern of target protein [7]. After getting the target protein, a common step is performing functional enrichment of these target proteins to get related biological processes and pathways of transcription factor or modified histone with R package such as clusterProfiler [8]. Motif analysis is widely used to identify binding patterns, and HOMER (http://homer.ucsd.edu/homer/motif/) can be used to discover motif de novo and screen for the enrichment of known motifs and visualize the sequence logo.

## 4 Notes

1. Given the small amount of mouse retina, 12 retinas are pooled for each ChIP reaction.
2. In the present protocol, we apply the single fixation method with formaldehyde to crosslink proteins to DNA in a reversible manner. The double fixation method described in the previous protocols [9], which need a second fixation with disuccinimidyl glutarate, is more suitable for investigating the proteins indirectly bound to DNA via protein-protein-DNA interaction, such as transcriptional cofactors.
3. Cryogenic grinding with liquid nitrogen is a widely used method to dissect tissues. However, a significant problem for cryogenic grinding is that a large amount of samples will lose during grinding, which leads to poor yields for retina tissues. Given that adhesions of retinal cells are weak, the retina tissues can be dissected easily by repeatedly blowing with pipette.

4. The glycine will be precipitated from 2.5 M glycine solution at 4 °C. Stock the 2.5 M glycine solution at room temperature.
5. The sonication conditions may need to be determined empirically by testing different sonicator power settings and sonication durations.
6. Except for the phenol/chloroform method, commercial DNA purification kits also work well for DNA purification.
7. Before sequencing, testing the enrichment of the positive region by PCR is necessary. Normally, choose two positive and negative regions for quality control to ensure that the ChIP reaction worked well.
8. To keep complexity of the sequencing library, we recommend using 5 ng ChIP-enriched DNA for transcription factors and 50 ng for modified histone proteins.

## References

1. Samuel A, Housset M, Fant B, Lamonerie T (2014) Otx2 ChIP-seq reveals unique and redundant functions in the mature mouse retina. PLoS One 9(2):e89110. https://doi.org/10.1371/journal.pone.0089110
2. Zibetti C, Liu S, Wan J, Qian J, Blackshaw S (2019) Epigenomic profiling of retinal progenitors reveals LHX2 is required for developmental regulation of open chromatin. Commun Biol 2: 142. https://doi.org/10.1038/s42003-019-0375-9
3. Popova EY, Barnstable CJ, Zhang SS (2014) Cell type-specific epigenetic signatures accompany late stages of mouse retina development. Adv Exp Med Biol 801:3–8. https://doi.org/10.1007/978-1-4614-3209-8_1
4. Tao Y, Cao J, Li M, Hoffmann B, Xu K, Chen J et al (2020) PAX6D instructs neural retinal specification from human embryonic stem cell-derived neuroectoderm. EMBO Rep 21(9): e50000. https://doi.org/10.15252/embr.202050000
5. Liang WJ, Yang HW, Liu HN, Qian W, Chen XL (2020) HMGB1 upregulates NF-kB by inhibiting IKB-alpha and associates with diabetic retinopathy. Life Sci 241:117146. https://doi.org/10.1016/j.lfs.2019.117146
6. Jorstad NL, Wilken MS, Todd L, Finkbeiner C, Nakamura P, Radulovich N et al (2020) STAT signaling modifies Ascl1 chromatin binding and limits neural regeneration from Muller glia in adult mouse retina. Cell Rep 30(7): 2195–2208. e5. https://doi.org/10.1016/j.celrep.2020.01.075
7. Yu G, Wang LG, He QY (2015) ChIPseeker: an R/Bioconductor package for ChIP peak annotation, comparison and visualization. Bioinformatics 31(14):2382–2383. https://doi.org/10.1093/bioinformatics/btv145
8. Wu T, Hu E, Xu S, Chen M, Guo P, Dai Z et al (2021) clusterProfiler 4.0: A universal enrichment tool for interpreting omics data. Innovation (N Y) 2(3):100141. https://doi.org/10.1016/j.xinn.2021.100141
9. Tian B, Yang J, Brasier AR (2012) Two-step cross-linking for analysis of protein-chromatin interactions. Methods Mol Biol 809:105–120. https://doi.org/10.1007/978-1-61779-376-9_7

Chapter 15

# Frozen Block Tissue Staining for Eye Structure of Zebrafish Embryo

**Yujin Lee and Jaewook Yang**


## Abstract

Histology is a technique used to study the morphology of cell structures by cutting samples into thin sections. Histological cross-section and staining are the techniques needed to visualize the morphology of cell tissues. A suitable tissue staining experiment was created to observe changes in the retinal layer in zebrafish embryos. Zebrafish have a human-like visual system, retina, and eye structures. Due to the small size of zebrafish and undeveloped bones in the embryonic stage, the resistance through cross-section is inevitably small. Here, we present optimized protocol changes in eye tissue of zebrafish using frozen blocks.


**Key words** Histology, Zebrafish, Embryo, Eye, Retina, Tissue, Frozen block

## 1 Introduction

Hematoxylin and eosin (H&E) stain is one of the main tissue staining methods used in histology [1]. H&E is the mixture of two histological stains. Hematoxylin stains cell nuclei, and eosin stains the extracellular matrix [2]. Therefore, it can easily differentiate between the nuclear and cytoplasmic parts of a cell. Additionally, the overall patterns show the general layout and provide a general overview of a tissue structure [3]. The H&E staining method is the main stain in histology, because it can be done quickly, is not expensive, and stains tissues in large amounts. Also, the H&E staining method is very routinely used in histology, because it is not dependent on the chemicals and protocols [4]. However, the common H&E staining procedure does not always fit all tissues. Therefore, more specific stains and methods are needed. In particular, pre-treatment step is necessary before conducting an experiment using zebrafish tissue because zebrafish live in water, unlike ordinary animals. Zebrafish is a very useful species in many different biomedical research areas. Zebrafish have genetic information, and many organs have similar structure

Guei-Sheung Liu and Jiang-Hui Wang (eds.), *Diabetic Retinopathy: Methods and Protocols*, Methods in Molecular Biology, vol. 2678, https://doi.org/10.1007/978-1-0716-3255-0_15,

as humans [5]. Zebrafish retinas are similar in morphology and physiology to human retinas. In particular, the retinal structure of zebrafish has five distinct layers [6–8]. After 96 h post-fertilization (hpf), zebrafish eye will have the same structure as the adult. Thus, the relationship between the ocular can be identified in zebrafish. A histological experiment is needed to observe zebrafish retinal layer changes [9]. Zebrafish are about 1 mm small in size at the embryonic stage, so they can be cross-sectionalized throughout the entire embryo [10]. In addition, the bones are not mature at the embryonic stage, so the tissue remains soft and has less resistance to amputation. Therefore, zebrafish offer an excellent animal model for studying ocular.

## 2 Materials

### 2.1 Tissue Processing and Staining

1. Zebrafish embryos are obtained through mating with adult zebrafish. Embryos are maintained in a suitable growth media in Petri dishes at 26–28 °C. We have used this protocol at 6 days post-fertilization (dpf) of zebrafish embryos (Fig. 1).
2. Formalin solution neutral buffered, 10% (equivalent 4% formaldehyde) (*see* **Note 1**).
3. Phosphate-buffered saline (PBS).
4. 30% sucrose in PBS.
5. Tissue-Tek® optimum cutting temperature (O.C.T.), embedding medium for frozen tissue.
6. Liquid nitrogen (*see* **Note 2**).

### 2.2 H&E Staining (Fig. 2; See Note 3)

1. Hematoxylin 450 mL (add acetic acid 4.5 mL) (*see* **Note 4**).
2. 1% eosin Y 450 mL (add acetic acid 4.5 mL).
3. 1% HCl: 70% ethanol 450 mL + HCl 4.5 mL.
4. 95% ethanol: 100% ethanol 427.5 mL + distilled water 22.5 mL.

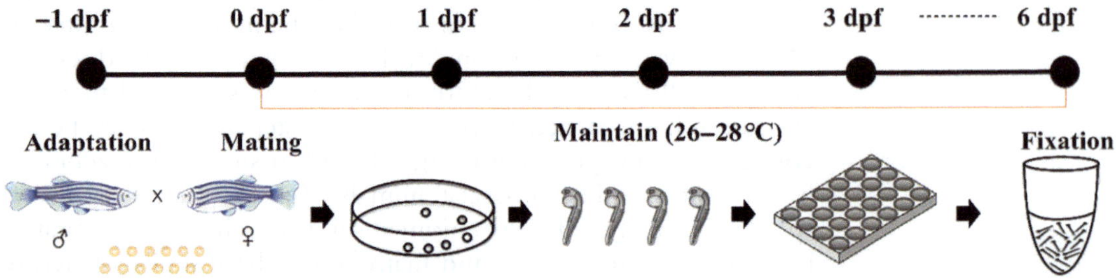


**Fig. 1** Adult zebrafish are maintained between 26 and 28 °C, and the lighting conditions are 14:10 h (light/dark). The zebrafish are raised in a stand-alone housing unit

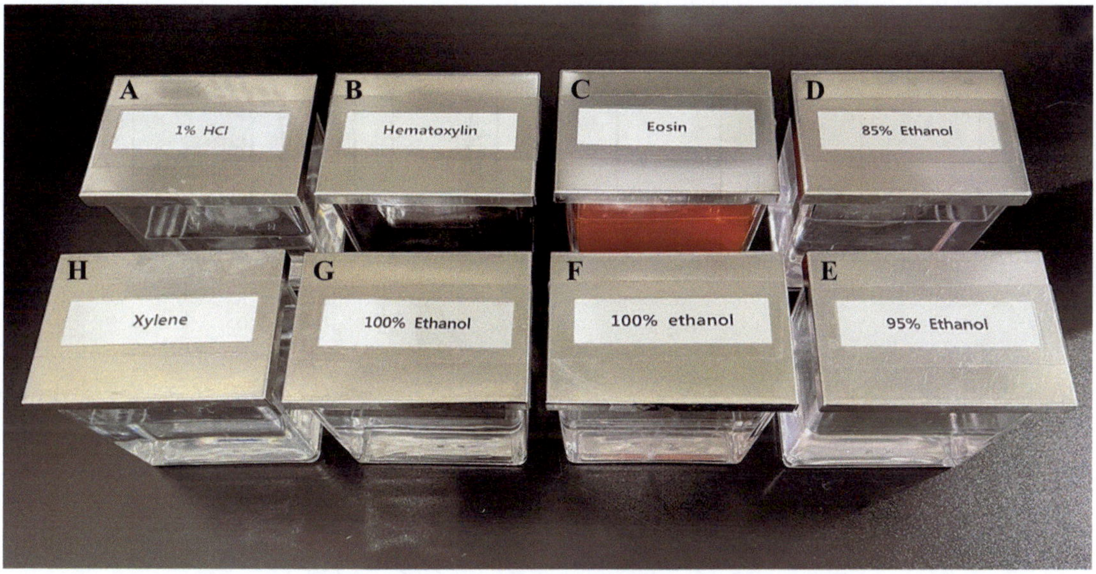


**Fig. 2** This is a picture of the prepared staining solution in a glass jar. Image shows 1% HCl (**a**), hematoxylin (**b**), eosin (**c**), 85% ethanol (**d**), 95% ethanol (**e**), 100% ethanol (**f**), 100% ethanol (**g**), and xylene (**h**)

5. 85% ethanol: 100% ethanol 382.5 mL + distilled water 67.5 mL.
6. 100% ethanol 450 mL.
7. Xylene 450 mL (*see* **Note 5**).
8. VectaMount permanent mounting solution.

## 3 Methods

### *3.1 Tissue Fixation*

The fixation stage is very important because it affects tissue staining results.

1. At 6 dpf, zebrafish embryos are placed in a 24-well plate (5–10 embryos/well).
2. The zebrafish embryos are fixed in formalin solution (1 mL/well) for 3–4 h at room temperature.
3. Remove the formalin solution and wash with PBS, and wash three times for 5 min each wash.
4. Remove the PBS and place zebrafish embryos in 30% sucrose diluted with PBS overnight until it sinks at 4 °C (*see* **Note 6**).
5. As shown in Fig. 3b, fill the disposable base mold full with O.C.T. compound (*see* **Note 7**).
6. Transfer the embryo to a disposable base mold with forceps (*see* **Note 8**).

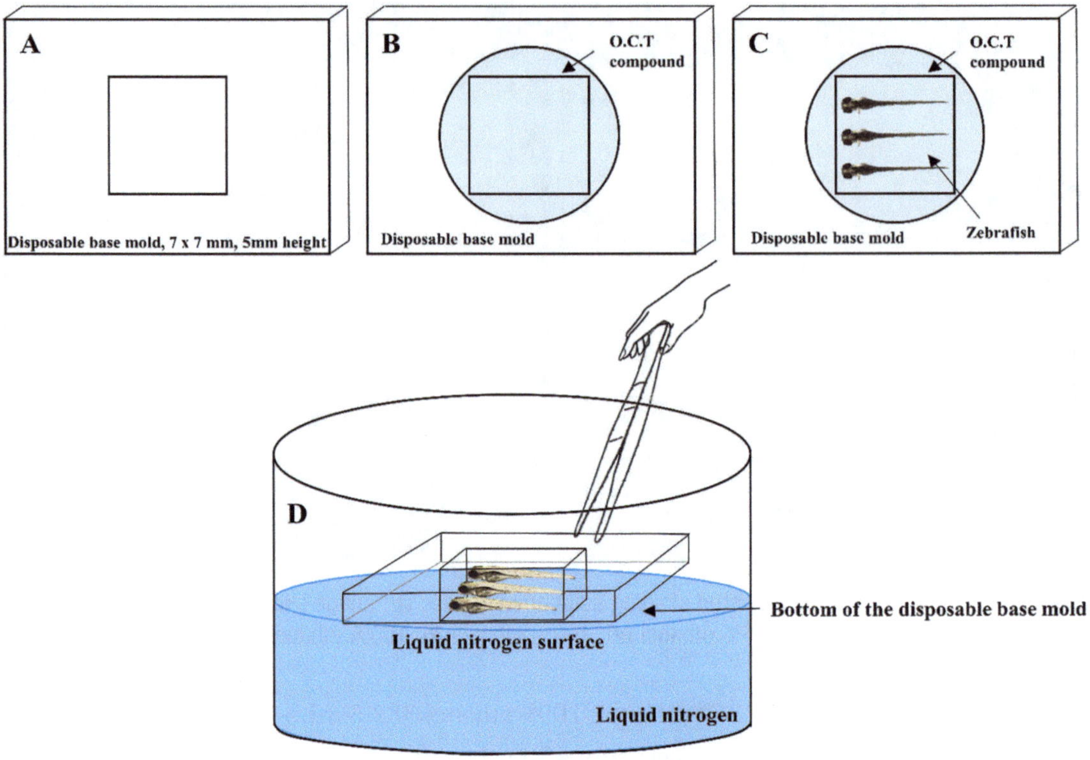


**Fig. 3** The procedure of embed zebrafish in a disposable base mold. Panel (**a**) is a disposable base mold and (**b**) shows the disposable base mold full of O.C.T. compound. Panel (**c**) shows embed zebrafish and (**d**) shows the freeze of zebrafish embryo tissue in liquid nitrogen

7. Embed zebrafish embryo in a disposable base mold containing O.C.T. compound (Fig. 3c) (*see* **Note 9**).
8. As shown in Fig. 2d, the disposable base mold is put in the liquid nitrogen for 3 min until the O.C.T. compound gets hard (Fig. 3d) (*see* **Note 10**).
9. Then, blocks are frozen overnight at −20 °C and sectioned at 8 μm on a cryostat (*see* **Note 11**).
10. Sections are collected on the microscope slides, and the slides are stored at −20 °C (*see* **Note 12**).

### 3.2 H&E Staining

1. Air-dry the microscope slide at room temperature for 30 min (*see* **Note 13**).
2. Place in the slide glass rack and wash with PBS for 10 min (*see* **Note 14**).
3. Place the slide glass rack into the hematoxylin solution for 1 min or 1 min 30 s (*see* **Note 15**).

4. Wash the slide glass rack with tap water (*see* **Note 16**).
5. Place the slide into a 1% HCl solution for three times (*see* **Note 17**).
6. Place the slide glass rack into the eosin solution for 1 min (*see* **Note 18**).
7. Wash the slide glass rack with tap water (*see* **Note 19**).
8. Place the slide glass rack into 85% ethanol solution for 3 min.
9. Place the slide glass rack into 95% ethanol solution for 3 min.
10. Place the slide glass rack into 100% ethanol solution for 3 min and 100% ethanol for 3 min in another glass jar.
11. Place the slide glass rack into xylene solution for 5 min.
12. Spray the mounting solution on the slide and cover it with a cover glass (*see* **Note 20**).
13. Air-dry the microscope slide at room temperature for 30 min or 1 h (*see* **Note 21**).
14. The following images are hematoxylin and eosin-stained sections using a virtual (Fig. 4).

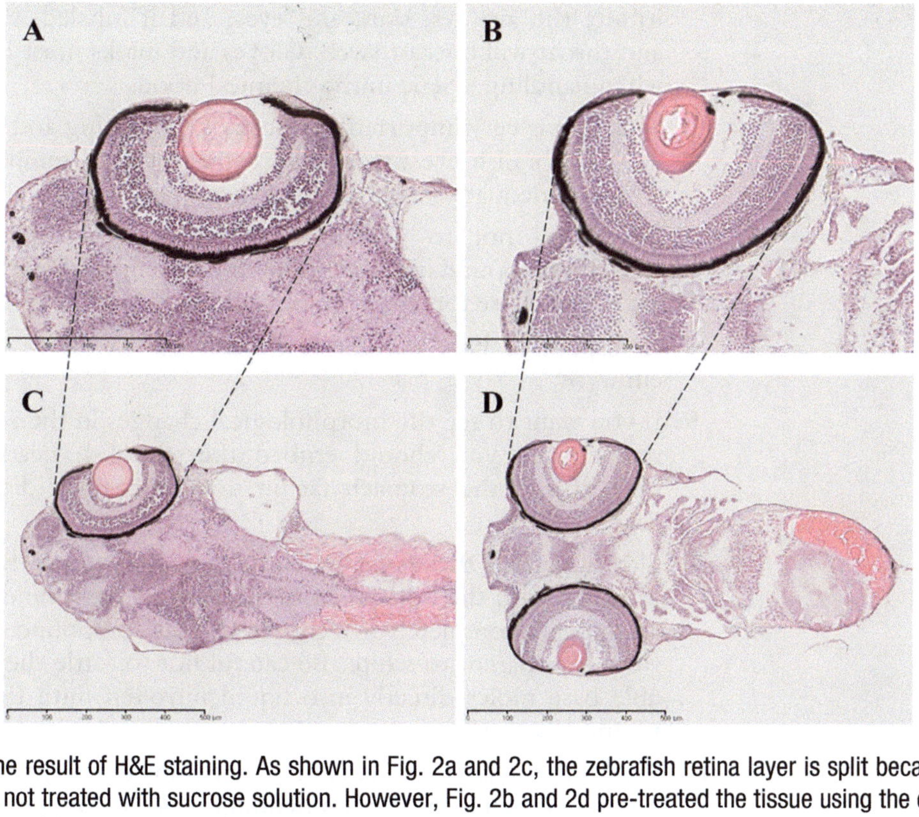


**Fig. 4** The result of H&E staining. As shown in Fig. 2a and 2c, the zebrafish retina layer is split because this tissue is not treated with sucrose solution. However, Fig. 2b and 2d pre-treated the tissue using the changed protocol, and it can observe the changes in the retina layer

## 4 Notes

1. Formalin is a carcinogen and toxic solution. Paraformaldehyde work must be done in the chemical hood, and appropriate personal protective equipment must always be used at all times. We typically buy a 10% solution and dilute it to 4% on PBS.
2. Liquefied nitrogen should be handled in a well-ventilated place. Only handle liquids slowly, and use approved containers that can withstand low temperatures. Personal protective equipment is also required. Wear safety glasses to protect your eyes and face. Safety shoes and leather gloves are recommended when handling containers.
3. Be careful when handling staining solutions. They will permanently stain your skin and clothes. Wear gloves and a mask while handling the solution in the chemical hood. Always keep the container closed when not in use. Never dispose of xylene, hematoxylin, and eosin in the sink.
4. Acetic acid is added to hematoxylin to make it more clear to stain.
5. Xylene is a colorless liquid that can cause dizziness. It can irritate the airways, skin, and eyes, and if inhaled, the nose and throat will hurt or swell. Gloves and masks must be worn when handling xylene in the chemical hood.
6. This step is very important because of the tearing and damaging of zebrafish eye tissue from extremely cold temperatures without adequate cryoprotection measures.
7. Be careful not to generate air bubbles when filling the O.C.T. compound into the disposable base mold. If air bubbles are present, use forceps to remove them.
8. A disposable base mold can contain up to three zebrafish embryos.
9. If you want to see the morphological changes in the zebrafish retina layer, you should embed the zebrafish head facing upward and the stomach facing downward in a disposable base mold.
10. Slowly harden the tissue by contacting the surface of liquid nitrogen with the bottom of the disposable base mold. When the block is completely frozen, the O.C.T. compound changes from transparent to white. Be careful not to settle the disposable base moles directly into liquid nitrogen until the compound hardens.
11. Freeze the tissue for 1 day until it is completely solidified. The section immediately after embedding is not recommended.

12. Freeze the tissue for approximately 1 day to allow it to adhere completely to the slide. Staining immediately after cutting is not recommended.
13. This step is to slowly melt frozen tissue. This step is to slowly melt frozen tissue. Leaving it at room temperature for at least 15 min is recommended because this process reduces tissue damage.
14. This step is to remove the O.C.T. compound attached to the slide and stain only the tissue.
15. We recommend hematoxylin staining time at 1 min if you prepared the solution for the first time. But, when you reuse the hematoxylin solution, we recommend staining it for 1 min 30 s.
16. This step is to remove excess stain until only the color remains on the tissue components.
17. The reason for this step is to bleach except for the nucleus. However, if the hematoxylin dye is light, this process may be omitted, or time may be shortened.
18. The staining time of eosin may vary depending on the staining time of hematoxylin. It is recommended that you soak it in a solution that comes in about 1/3 of the staining time of hematoxylin.
19. This step is required to remove any remaining dye reagents on the slide.
20. Make sure that the cover glass and slides are properly attached. Otherwise, it is difficult to analyze the tissue during scanning because air can enter the tissue.
21. This step is intended to prevent the tissue from being pushed or damaged during installation.

## Acknowledgments

This work was supported by EYEBIOKOREA Inc. and Inje University Busan Paik Hospital.

### References

1. Titford M (2005) The long history of hematoxylin. Biotech Histochem 80:73–78
2. Chan JK (2014) The wonderful colors of the hematoxylin-eosin stain in diagnostic surgical pathology. Int J Surg Pathol 22(1):12–32. https://doi.org/10.1177/1066896913517939
3. Wittekind D (2003) Traditional staining for routine diagnostic pathology including the role of tannic acid. 1. Value and limitations of the hematoxylin-eosin stain. Biotech Histochem 78:261–270
4. Schulte EK (1991) Standardization of biological dyes and stains: pitfalls and possibilities. Histochemistry 95(4):319–328. https://doi.org/10.1007/BF00266958
5. Ota S, Kawahara A (2014) Zebrafish: a model vertebrate suitable for the analysis of human

genetic disorders. Congenit Anom (Kyoto) 54(1):8–11. https://doi.org/10.1111/cga.12040
6. Bibliowicz J, Tittle RK, Gross JM (2011) Toward a better understanding of human eye disease insights from the zebrafish, Danio rerio. Prog Mol Biol Transl Sci 100:287–330
7. Chhetri J, Jacobson G, Gueven N (2014) Zebrafish—on the move towards ophthalmological research. Eye 28:367–380
8. Wan J, Goldman D (2016) Retina regeneration in zebrafish. Curr Opin Genet Dev 40:41–47
9. Li H, Yu S, Cao F et al (2018) Developmental toxicity and potential mechanisms of pyraoxystrobin to zebrafish (Danio rerio). Ecotoxicol Environ Saf 151:1–9
10. Sullivan-Brown J, Bisher ME, Burdine RD (2011) Embedding, serial sectioning and staining of zebrafish embryos using JB-4 resin. Nat Protoc 6(1):46–55. https://doi.org/10.1038/nprot.2010.165

# Chapter 16

# Investigating the Function of MicroRNAs in Human Retinal Microvascular Endothelial Cells of Diabetic Retinopathy

**Yong Zeng, Zekai Cui, Jiansu Chen, and Shibo Tang**

## Abstract

Diabetic retinopathy (DR) is the main complication of diabetes mellitus (DM). Recent studies have implicated microRNAs dysfunction in human retinal microvascular endothelial cell (HRMEC). In this study, we aim to investigate the apoptotic promotion of miR-29b-3p by blocking SIRT1 in HRMEC for DR situation. To identify the regulating relationship between miR-29b-3p and SIRT1, HRMECs were transfected with miR-29b-3p mimics/inhibitors or their negative controls. Cell viability was assessed with the cell counting kit-8 (CCK-8) assay, and apoptotic cells were stained by one-step TUNEL assay kit. Gene and protein expression were assayed by RT-qPCR and Western blotting separately. Dual-luciferase reporter assay using HEK293T cells was performed to show the direct interaction of miR-29b-3p and the 3′-UTR of SIRT1. HRMECs were identified as >95% positive for CD31 and vWF. Upregulated miR-29b-3p decreased the expression of SIRT1 and increased the ratio of Bax/Bcl-2, while downregulated miR-29b-3p increased the expression of SIRT1 protein and downregulated the ratio of Bax/Bcl-2. Dual-luciferase reporter assay showed the direct interaction of miR-29b-3p and SIRT1. The dysregulation of miR-29b-3p/SIRT1 is a potential mechanism of HRMEC apoptosis in DR. miR-29b-3p/SIRT1 may be a potential therapeutic target for DR.

**Key words** Diabetic retinopathy, miR-29b-3p, HRMEC, Apoptosis

## 1 Introduction

MicroRNA is a type of noncoding RNA composed of 19–25 nucleotides [1]. Thousands of microRNAs have been discovered since they were first reported in 1993, and a single microRNA may have hundreds of target mRNAs [2, 3]. MicroRNA regulates post-transcriptional gene expression via binding to target sites directly or promoting mRNA degradation [3]. A series of studies have revealed the regulation of microRNA in aging, tumor progression, metabolic diseases, and inflammation [4–7]. In recent years, a variety of microRNAs (miR-409-3p, miR-98-5p, miR-16-5p, etc.) are proven to participate in diabetes progression and its

Guei-Sheung Liu and Jiang-Hui Wang (eds.), *Diabetic Retinopathy: Methods and Protocols*, Methods in Molecular Biology, vol. 2678, https://doi.org/10.1007/978-1-0716-3255-0_16,

complications [8–10]. Furthermore, numerous microRNAs (miR-3197, miR-2116-5p, miR-152, miR-34a, etc.) are identified as specificity biomarkers, and they participate directly in DR progression [11–13].

Recent studies have revealed the decrease of SIRT1 in diabetes patients [14] and the increase of miR-29b-3p in diabetes [15]. Furthermore, Su et al. have proved the direct regulation of miR-29b-3p to SIRT1 in insulin resistance [16]. To our knowledge, there is no research to illuminate the regulatory relationship of the miR-29b-3p/SIRT1/Bax/Bcl-2 pathway in human retinal microvascular endothelial cells (HRMECs). Here, we perform this study to verify the role of miR-29b-3p in DR.

## 2 Materials

### 2.1 Adult HRMEC Preparation

1. Adult donor eye.
2. Dissecting microscope.
3. Ophthalmic tweezers and scissors.
4. 0.25% EDTA-trypsinase and 0.1% type II collagenase.
5. Endothelial complete medium: ECM containing 5% fetal bovine serum and 1% penicillin-streptomycin (P/S).
6. 100 μm filter and T-25 cm$^2$ flask precoated with 1% gelatin.
7. von Willebrand factor (vWF) and CD31 antibodies.

### 2.2 HRMEC Transfection

The miR-29b-3p mimics, inhibitors, negative controls (NC), and transfection reagent were purchased commercially.

### 2.3 Cell Apoptotic and Viability Assay

1. TUNEL apoptosis assay kit.
2. 4′,6-Diamidino-2-phenylindole (DAPI).
3. Cell counting kit-8 (CCK-8) assay kit.

### 2.4 qRT-PCRs

1. MicroRNA assay kit (*see* **Note 1**).
2. Reverse transcription kit.

### 2.5 Dual-Luciferase Reporter Assay

1. HEK293T cell line.
2. The 3′-UTR predicted binding site vectors were purchased commercially.
3. Dual-luciferase assay system.

## 3 Methods

### 3.1 Cell Culture

1. Retinas isolated from the eyes are minced into small pieces (*see* **Note 2**).
2. The retinas are washed with cold PBS containing 5% penicillin/streptomycin. The fragments are retained on a 100 μm cell strainer.
3. Following harvesting, the fragments are transferred from the sieve to a 15 mL tube containing 3 mL 0.1% type II collagenase at 37 °C for 30 min with agitation and then added 3 mL ECM to terminate digestion. Centrifuge at 300×*g* for 10 min, and wash the pellet with ECM once.
4. The retinal fragment suspensions are filtered through a 100 μm cell strainer.
5. After centrifugation at 300×*g* for 10 min, the pellets containing microvessel fragments are resuspended in endothelial cell medium. The above suspension is transferred to a T-25 $cm^2$ flask precoated with 1% gelatin, maintained in the ECM, and cultured at 37 °C in a 5% $CO_2$ humidified incubator.
6. The formation of cells is checked daily; after 48 h, 50% of the medium was renewed, and thereafter, the medium is changed every 2 days. The confluent cultures are passaged by detaching the cells using 0.1% trypsin in PBS for 2–3 min and plating at a split ratio of 1:2. The cells used in this study are passaged three to five times and stained with primary antibodies recognizing vWF and CD31. The images are captured using a fluorescence microscope (Fig. 1).

### 3.2 Cell Transfection

1. Cells are seeded in 6-well and 96-well plates with a density of $2 \times 10^5$/well and $4 \times 10^3$/well (*see* **Notes 3** and **4**).
2. The miR-29b-3p mimics, inhibitors, and their negative controls (NC) are transfected into cells using a transfection reagent accordingly.

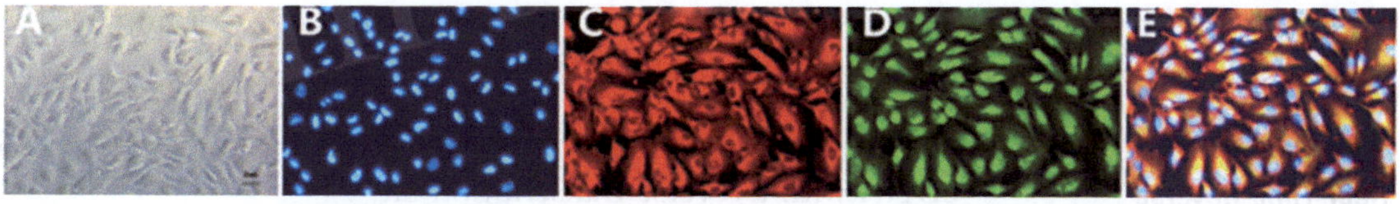


**Fig. 1** Culture and verification of HRMEC. (**a**) After 10 days of culture, the cells showed an oval morphology and a contact-inhibited monolayer. (**b**) Cell nuclei were stained with DAPI (blue fluorescent). (**c**) CD31 was positive (red fluorescent). (**d**) vWF was positive (green fluorescent). (**e**) The merged picture revealed CD31/vWF double staining positive. The cell morphology and immunofluorescent results verified the HRMEC cell type. HRMEC, human retinal microvascular endothelial cell; CD31, platelet endothelial cell adhesion molecule-1; vWF, von Willebrand factor

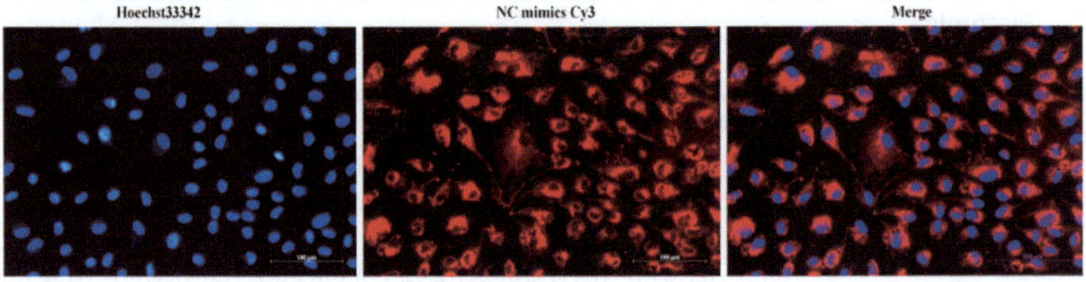


**Fig. 2** miR-29b-3p mimics, inhibitors, and their negative controls were transfected into HRMECs successfully. Annulus red fluorescence was observed around the nuclear after 30 h's transfection with NC mimics Cy3

3. NC mimics labeled with Cy3 fluorescence (Guangzhou, China) are transfected to observe the transfection efficiency directly (Fig. 2).
4. After 30 h of transfection, the HRMECs are collected for TUNEL stain, cell counting kit-8 (CCK-8) assay, and qRT-PCR.

### 3.3 Cell Apoptotic and Viability Assay

1. $4 \times 10^3$ cells/well are seeded into 96-well plates and cultured at 37 °C with 5% $CO_2$ in a humidified environment.
2. The one-step TUNEL apoptosis assay is used for detecting apoptotic cells.
3. Nuclei is stained by DAPI (blue). Fluorescent images are acquired by a fluorescence microscope.
4. The quantification of TUNEL-positive cells is obtained by ImageJ software and calculated by GraphPad Prism version 5.0.
5. Cell viability is determined by a CCK-8 assay.
6. Seven replicates per group and a group without cells served as the blank.
7. After treatment with different conditions, 100 μL of fresh culture medium with 10% CCK-8 solution is added to each well and incubated at 37 °C for 1.5 h.
8. The absorbance at 450 nm is observed by Synergy™ HTX Multi-Mode Microplate Reader.
9. The relative viability of cells is calculated according to the manufacturer's protocol.

### 3.4 Quantitative Real-Time Reverse Transcriptase-Polymerase Chain Reactions (qRT-PCRs)

1. MicroRNA is isolated with a microRNA kit and reversed to cDNA with a reverse transcription kit.
2. The RT product is subjected to 45 cycles of qPCR reactions.
3. U6 is used to normalize the expression of microRNA. The relative expression level of miRNA was calculated by the $2^{-\Delta\Delta CT}$ method.
4. The specific primers for miR-29b-3p and U6 are listed in Table 1 (*see* **Note 5**).

**Table 1**
**The sequences of specific primers**

| Human genes | Primer sequences | |
|---|---|---|
| miR-29b-3p | Stem-loop primer | GTCGTATCCAGTGCAGGGTCCGAGGTATTCGCACTGGATACGACAACACTGA |
| | Forward primer | CTGCTAGCACCATTTGAAA |
| | Reverse primer | GTGCAGGGTCCGAGGT |
| U6 | Forward primer | CTCGCTTCGGCAGCACA |
| | Reverse primer | AACGCTTCACGAATTTGCGT |

### *3.5 Dual-Luciferase Reporter Assay*

1. 293T cells are cultured at 37 °C in a 5% $CO_2$ humidified incubator.
2. 293T cells are inoculated into 96-well plates with $1.5 \times 10^4$ cells per well, and the total volume of each well is 100 μL.
3. Take 10 μL ECM to dilute miRNA mimics or non-target control.
4. Take 15 μL ECM to dilute 3'-UTR double reporter vector or mutant vector.
5. The 0.25 μL transfection reagent is diluted in 25 μL ECM medium. After 5 min, the three reagents are mixed with a total of 50 μL, gently shaken, and left for 20 min.
5. Before adding plasmids and mimics into cells, each well is siphoned off 50 μL medium, and then the above 50 μL mixture was added. Finally, the total volume of each well is 100 μL. The mimics transfection concentration is 50 nM, and the plasmid concentration is 100 ng/well. Each group had three multiple wells. After 6 h, 100 μL fresh medium is added.
6. Before fluorescence detection, the luciferase substrate and luciferase buffer of the kit are mixed and divided, stored at −80 °C, and balanced to room temperature before use. The Stop & Glo buffer is divided and stored at −80 °C and balanced to room temperature before use, and an appropriate amount of Stop & Glo substrate was added.
7. After transfection for 48 h, the medium is aspirated and fresh medium is added with 35 μL/well, and the luciferase substrate is added with 35 μL/well, and the fluorescence value is measured after shaking for 10 min.
8. Add 30 μL Stop & Glo reagent, oscillate for 10 min, and determine the fluorescence value by fluorescence illuminometer (Fig. 3).

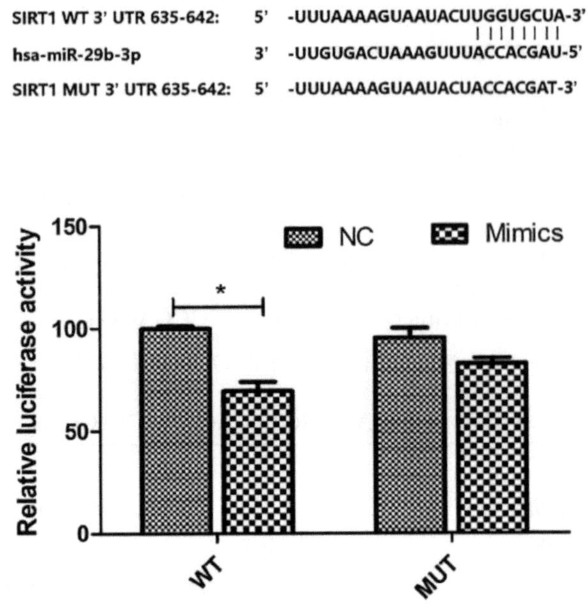


**Fig. 3** The dual-luciferase reporter assay of miR-29b-3p and SIRT1. SIRT1 is a direct target of miR-29b-3p predicted by the miRNA online database (TargetScanHuman7.2 and miRBase). Dual-luciferase reporter assay using HEK293T cells showed the direct interaction of miR-29b-3p and the 3′-UTR of SIRT1. After 48 h of cotransfection, overexpression of miR-29b-3p reduced the luciferase activity of the WT reporter but had no inhibition on the MUT reporter. Data were shown as mean ± SD, *$p<0.05$, *NC* negative control, *Mimics* miR-29b-3p mimic, *WT* wild type, *MUT* mutant type

## 4 Notes

1. An appropriate microRNA isolation kit is proposed; we select a commercial kit (Omega).
2. Retina tissue should be as fresh as possible for a successful HRMEC isolation.
3. PBMCS are seeded into culture plates with a proper density for a few hours of incubation in the morning and transfected in the afternoon.
4. Cells should be starved for 6–8 h before transfecting to avoid the affection of plasma factor and penicillin-streptomycin.
5. The stem-loop method is especially used for microRNA reverse transcription.

## References


1. Lu TX, Rothenberg ME (2018) MicroRNA. J Allergy Clin Immunol 141(4):1202–1207
2. O'Kelly F et al (2012) MicroRNAs as putative mediators of treatment response in prostate cancer. Nat Rev Urol 9(7):397–407
3. Landrier JF et al (2019) MicroRNAs in obesity and related metabolic disorders. Cell 8(8):859
4. Klieser E et al (2019) The crosstalk of miRNA and oxidative stress in the liver: from physiology to pathology and clinical implications. Int J Mol Sci 20(21):5266
5. Majidinia M et al (2019) Targeting miRNAs by polyphenols: novel therapeutic strategy for aging. Biochemist Pharmacol 173:113688
6. Nasr MA et al (2019) Dysregulated MicroRNA fingerprints and methylation patterns in hepatocellular carcinoma, cancer stem cells, and mesenchymal stem cells. Front Cell Dev Biol 7:229
7. Zhao M et al (2019) The regulatory role of non-coding RNAs on programmed cell death four in inflammation and cancer. Front Oncol 9:919
8. Duan YR et al (2019) Exosomal microRNA-16-5p from human urine-derived stem cells ameliorates diabetic nephropathy through protection of podocyte. J Cell Mol Med 25(23): 10798–10813
9. Khan R et al (2019) Circulatory miR-98-5p levels are deregulated during diabetes and it inhibits proliferation and promotes apoptosis by targeting PPP1R15B in keratinocytes. RNA Biol 17:1–14
10. Ventriglia G et al (2019) miR-409-3p is reduced in plasma and islet immune infiltrates of NOD diabetic mice and is differentially expressed in people with type 1 diabetes. Diabetologia 63:124
11. Fu X, Ou B (2019) miR-152/LIN28B axis modulates high-glucose-induced angiogenesis in human retinal endothelial cells via VEGF signaling. J Cell Biochem 121(2):954–962
12. Ji H et al (2019) Circulating miR-3197 and miR-2116-5p as novel biomarkers for diabetic retinopathy. Clin Chim Acta 501:147–153
13. Thounaojam MC et al (2019) MicroRNA-34a (miR-34a) mediates retinal endothelial cell premature senescence through mitochondrial dysfunction and loss of antioxidant activities. Antioxidants (Basel) 8(9):328
14. Balestrieri ML et al (2013) Poor glycaemic control in type 2 diabetes patients reduces endothelial progenitor cell number by influencing SIRT1 signalling via platelet-activating factor receptor activation. Diabetologia 56(1): 162–172
15. Esteves JV et al (2018) Diabetes modulates MicroRNAs 29b-3p, 29c-3p, 199a-5p and 532-3p expression in muscle: possible role in GLUT4 and HK2 repression. Front Endocrinol (Lausanne) 9:536
16. Su T et al (2019) Bone marrow mesenchymal stem cells-derived Exosomal MiR-29b-3p regulates aging-associated insulin resistance. ACS Nano 13(2):2450–2462

# Chapter 17

# Genome Editing Inhibits Retinal Angiogenesis in a Mouse Model of Oxygen-Induced Retinopathy

**Wenyi Wu and Hetian Lei**


## Abstract

This protocol describes a novel approach harnessing the technology of clustered regularly interspaced short palindromic repeats (CRISPR)/CRISPR-associated (Cas) 9-based gene editing for treating retinal angiogenesis. In this system, adeno-associated virus (AAV)-mediated CRISPR/Cas9 was employed to edit the genome of vascular endothelial growth factor receptor (VEGFR)2 in retinal vascular endothelial cells in a mouse model of oxygen-induced retinopathy. The results showed that genome editing of *VEGFR2* suppressed pathological retinal angiogenesis. This mouse model mimics a critical aspect of abnormal retinal angiogenesis in patients with neovascular diabetic retinopathy and retinopathy of prematurity, indicating genome editing has high potential for treating angiogenesis-associated retinopathies.


**Key words** CRISPR/Cas9, Vascular endothelial cells, Angiogenesis, Oxygen-induced retinopathy

## 1 Introduction

Diabetic retinopathy ranks as the leading cause of blindness and visual loss of the working class worldwide [1, 2]. With the prevalence of anti-vascular endothelial growth factor (VEGF) therapy [3, 4], this effective treatment needs repeat intravitreal injection, and poor compliance can lead to disappointing effects. Thus, development of new treatment strategies that provide more durable treatment effects could help address some of these unmet needs. Gene-editing therapeutics are very promising to be used in retinal disease due to its local treatment such as diabetic retinopathy. Different gene-based therapeutic strategies have been attempted, including gene augmentation therapy to induce in vivo expression of therapeutic molecules and gene editing to knock down genes encoding specific mediators in disease-associated pathways [5].

Adeno-associated viruses (AAVs) are generally considered at lower biosafety levels and elicit relatively low immunological effects in vivo; however, the confined capacity of AAV restricts its

Guei-Sheung Liu and Jiang-Hui Wang (eds.), *Diabetic Retinopathy: Methods and Protocols*, Methods in Molecular Biology, vol. 2678, https://doi.org/10.1007/978-1-0716-3255-0_17,

application in gene therapy. Thus, dual AAV system with SpCas9 and SpGuide or using smaller Cas protein can solve this problem, but the efficiency of smaller Cas9 such as SaCas9 or CjCas9 is lower than that of SpCas9 [6, 7]. Thereby, the dual AAV system with SpCas9 and SpGuide was utilized in this protocol [8, 9]. Here, we describe an efficient in vivo genome editing method of delivering CRISPR-Cas9 into vascular endothelial cells with AAVs to deplete VEGF receptor (VEGFR)2 for blocking pathological retinal angiogenesis in a mouse model of oxygen-induced retinopathy (OIR) [8].

In this protocol, expression of SpCas9 is driven by a specific endothelial promoter of intercellular adhesion molecule 2 (pICAM2) to restrict this foreign enzyme in vascular endothelial cells, which can be efficiently infected by AAV1. We exemplify this approach by editing *VEGFR2* in retinal vascular endothelial cells in a mouse model of oxygen-induced retinopathy and expect that this simplified protocol can be expanded to other editing targets in diabetic retinopathy (Fig. 1) [8, 10].

## 2 Materials

### 2.1 Reagents

1. *BsmBI* restriction enzyme.
2. DMEM (1×), with 4.5 g L-/D-glucose and L-glutamine, without sodium pyruvate.
3. EBM-2 endothelial cell growth basal medium.
4. Fetal bovine serum (FBS).
5. Lipofectamine 3000 transfection reagent.
6. Opti-MEM I reduced serum medium.
7. QuickExtract DNA Extraction Solution.
8. Quick-gDNA MidiPrep.
9. Penicillin-streptomycin, solution stabilized, with 10,000 units penicillin and 10 mg streptomycin/ml, sterile-filtered, BioReagent, suitable for cell culture.
10. Polybrene or hexadimethrine bromide.
11. Puromycin dihydrochloride.
12. SapI restriction enzyme.
13. T4 DNA ligase.
14. Polyethylenimine.
15. Pluronic F-68.
16. DNase l.
17. RNase A.
18. Amicon Ultra-15 Centrifugal Filter Unit.

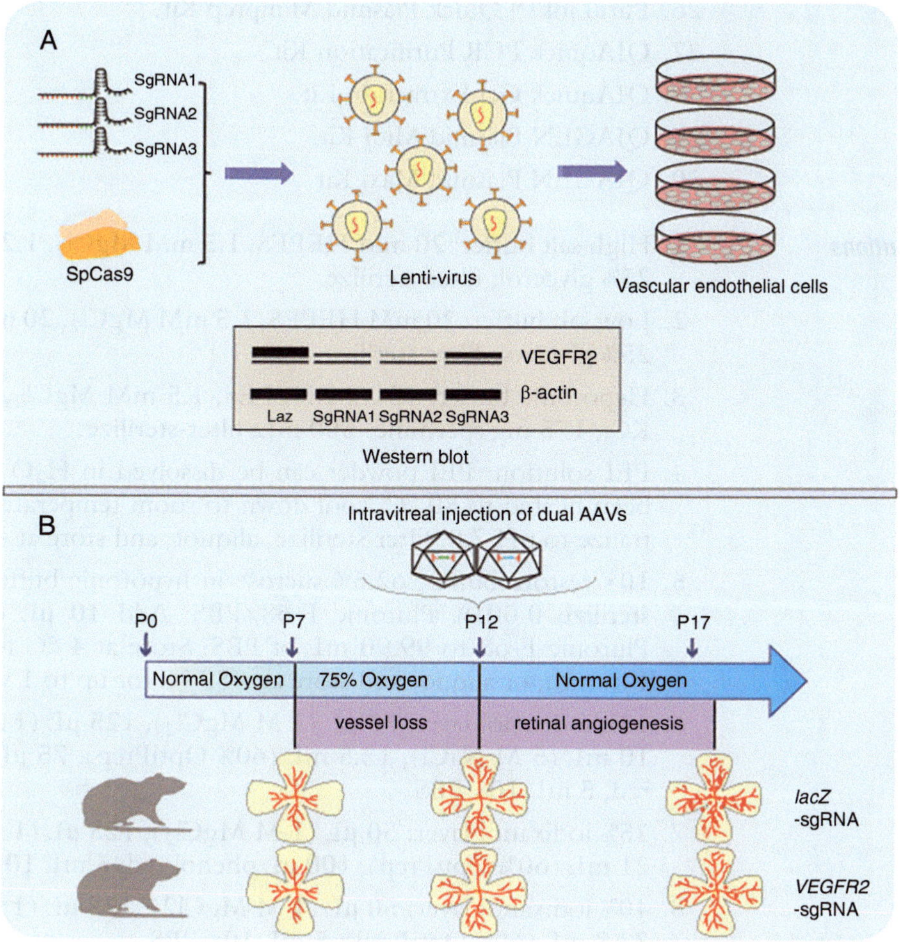


**Fig. 1** Editing *VEGFR2* inhibiting angiogenesis in a mouse model of OIR. (**a**) Selection of an effective sgRNA from three chosen sgRNAs for editing VEGFR2 in cultured vascular endothelial cells mediating using a lentiCRISPR/SpCas9. The editing efficiency was determined by Sanger sequencing and Western blot analysis. (**b**) Dual AAVs of SpCas9 and SpGuide were injected intravitreally in postnatal 12-day-old (P12) mice which had experienced in a high-oxygen (75%) chamber for 5 days. The mice were euthanized on P12 for future analyses including next-generation sequencing, Western blot, and isolectin B4 staining on the whole mount retina. These experiments were harnessed to determine the VEGFR2-editing efficiency and effects on retinal angiogenesis. Control AAV-pICAM2-SpCas9 and AAV-CMV-*lacZ*-sgRNA; *VEGFR2*, AAV-pICAM2-SpCas9 and AAV-CMV-*VEGFR2*-sgRNA

19. OptiSeal™ Polypropylene Tube.
20. Herculase II fusion DNA polymerase.
21. LB broth, ready-made powder.
22. Agar.
23. Carbenicillin disodium salt.
24. 5-Alpha competent *Escherichia coli*.
25. SOC outgrowth medium.

26. PureLink™ Quick Plasmid Miniprep Kit.
27. QIAquick PCR Purification Kit.
28. QIAquick Gel Extraction Kit.
29. QIAGEN Plasmid Midi Kit.
30. QIAGEN Plasmid Maxi Kit.

### 2.2 Solutions

1. High salt buffer: 20 mM HEPES, 1.5 mM $MgCl_2$, 1.2 M KCl, 25% glycerol; filter-sterilize.
2. Low salt buffer: 20 mM HEPES, 1.5 mM $MgCl_2$, 20 mM KCl, 25% glycerol; filter-sterilize.
3. Hypotonic buffer: 10 mM HEPES, 1.5 mM $MgCl_2$, 10 mM KCl, 175 mg spermine/500 mL; filter-sterilize.
4. PEI solution: PEI powder can be dissolved in $H_2O$ that has been heated to 80 °C, cool down to room temperature, neutralize to pH 7.0, filter-sterilize, aliquot, and store at −20 °C.
5. 10× restore buffer: 62.5% sucrose in hypotonic buffer; filter-sterilize 0.001% Pluronic F-68/PBS: Add 10 μL of 100× Pluronic F-68 to 99.90 mL of PBS. Store at 4 °C for up to 1 month, or aliquot and store at −80 °C for up to 1 year.
6. 15% iodixanol layer: 50 μL (1 M $MgCl_2$), 125 μL (1 M KCl), 10 mL (5 M NaCl), 12.5 mL (60% OptiPrep), 75 μL phenol red, 5 mL 10× PBS.
7. 25% iodixanol layer: 50 μL (1 M $MgCl_2$), 125 μL (1 M KCl), 21 mL (60% OptiPrep), 100 μL phenol red, 5 mL 10× PBS.
8. 40% iodixanol layer: 50 μL (1 M MgCl2), 125 μL (1 M KCl), 33.3 mL (60% OptiPrep), 5 mL 10× PBS.
9. 60% iodixanol layer: 50 μL (1 M MgCl2), 125 μL (1 M KCl), 50 mL (60% OptiPrep), 25 μL phenol red.

### 2.3 Mammalian Cells

1. Human retinal pigment epithelial cells (ARPE-19).
2. Human embryo kidney (HEK) 293T.
3. Primary human retinal microvascular endothelial cells.
4. Primary mouse brain vascular endothelial cells.

### 2.4 Plasmids

1. pAAV-U6-sgRNA-CMV-GFP (Addgene).
2. pAAV-pICAM2-SpCas9 (Addgene).
3. pAAV-RSV-SpCas9 (Addgene).
4. pAAV-RC plasmid for AAV1 replication and capsid genes (Addgene).
5. Helper plasmid for carrying adenovirus-derived genes (Addgene).

## 3 Methods

### 3.1 Selection of sgRNA for Editing VEGFR2

Vascular endothelial growth factor receptor (VEGFR)2 here is selected as a model molecule for preventing angiogenesis in a mouse model of oxygen-induced retinopathy (ROP) based on previous publications [8]. For genome editing-mediated depletion of *VEGFR2*, *Streptococcus pyogenes* (Sp)Cas9 is chosen due to its high efficiency, and its single guide (sg)RNAs can be identified at the website (https://chopchop.cbu.uib.no) from exons 2 to 5.

AAV-SpGuide together with AAV-pICAM2-SpCas9 (pICAM2 as an endothelial specific promoter/intercellular adhesion molecule 2) derived from the plasmid is transfected into vascular endothelial cells by electroporation for evaluating their efficiency in depleting VEGFR2 expression. Then the best-efficient sgRNAs in an AAV-SpGuide and AAV-pICAM2-SpCas9 were used to produce AAV1 in 293T cells, which can efficiently transduce vascular endothelial cells [8].

### 3.2 Cloning of sgRNAs into an AAV Vector to Produce AAV

1. Annealing and ligation
   (a) Three selected sgRNAs were cloned into AAV-SpGuide plasmid, respectively.
   (b) AAV-SpGuide plasmid was first digested by *SapI* restriction enzyme (Table 1). Incubate at 37 °C for 2 h, separate digested DNA fragments in a 1% agarose gel, and isolate the digested vector using QIAquick Gel Extraction Kit.
   (c) Heat 600–700 mL of water in a liter beaker on a heater until the water is boiling. Drop the tube with equal amounts of synthesized top and bottom strands (Table 2) on a floater into the water, and keep boiling for 5 min. Then cool it down gradually (*see* **Note 1**).
   (d) Incubate annealed oligos with purified DNA fragments from a *SapI*-digested AAV-SpGuide plasmid at room temperature (22 °C) for 2 h or at 16 °C overnight (Table 3) (*see* **Note 2**).

**Table 1**
**Preparation of vector digestion**

| Component | Amount per reaction |
|---|---|
| Plasmid | X μL |
| *SapI* or Bmbs | 2 μL |
| 10× CutSmart Buffer | 10 μL |
| $ddH_2O$ | Y μL |
| Total | 100 μL |

**Table 2**
**Preparation of sgRNA oligos annealing**

| Component | Amount per reaction |
|---|---|
| Top strand | 10 μM 10 μL |
| Bottom strand | 10 μM 10 μL |

**Table 3**
**Preparation of ligation reaction**

| Component | Amount per reaction |
|---|---|
| Annealed oligos (1 μg/mL each) | 1 μL |
| 10 × T4 DNA ligase buffer | 1 μL |
| T4 DNA ligase | 1 μL |
| Purified DNA fragment (30–60 ng)<br>$ddH_2O$<br>Total | X μL<br>Y μL<br>10 μL |

2. Transformation

   Thaw competent *E. coli* Stbl3 on ice slowly, transfer 2 ~ 5 μL of the ligation into Stbl3 (50 ~ 100 μL) competent cells and mix them gently, incubate the mixture on ice for 30 min, heat pulse for 90 s at 42 °C, let them stay on ice for 3 ~ 5 min, add LB 50 μL into the heat-shocked mixture, plate these bacteria onto a LB agar with ampicillin evenly, and then incubate the inverted plates at 37 °C overnight (about 16 h).

3. Sanger DNA sequencing

   Pick up 2 ~ 3 colonies from each transformation and culture them in LB with ampicillin overnight, prepare mini-preps, and then sequence isolated plasmids with a primer, 5′--GAGGGCCTATTTCCCATG-3′. This oligo sequence is from the U6 promoter for driving the transcription of sgRNA in the AAV-SpGuide plasmid.

4. Electroporation and confirmation of genome editing and protein depletion

   Transfer the dual AAV vectors (SpCas9 and SpGuide) into mouse brain microvascular endothelial cells by electroporation. Three days later, DNA fragments around protospacer adjacent motif (PAM) about 200 bp from genomic DNA transduced by the dual AAV vectors were PCR amplified for Sanger DNA sequencing to verify if there are indels that appeared at expected sites; in addition, Western blot can be used to verify whether there is a VEGFR2 depletion in the transfected vascular endothelial cells (Fig. 1).

5. Production of AAV virus
   (a) On the first day, plate HEK293T cells on ten 15 cm dishes for plasmid transfection (*see* **Note 3**).
   (b) On the second day, change medium into 20 mL freshly warmed up medium before transfection. Prepare DNA solution into a 50 mL falcon tube. The amounts of DNA and reagents per dish are shown below in the table (*see* **Notes 4** and **5**). Based on the size of ITR plasmid, calculate the amount of DNA carefully. Prepare one master mixture tube for five 15 cm dishes (all plasmids are diluted to 1 μg/μL in sterile $H_2O$). Prepare 10 mL DMEM (without phenol red) plus 785 μL PEI, mix again, and incubate at room temperature for 20–-30 minutes (*see* **Note 6**).
   (c) Add 2 mL of DNA-PEI mixture into 15 cm dish (totally ten dishes), and incubate the transfected cells.
   (d) On the third day, remove the medium, add 25 mL freshly warmed up medium, and let the cell grow for 2 days.
   (e) On the fifth day, scrape cells with cell scraper in their current medium, and transfer to 50 mL tubes. Spin at 2000×*g* for 5 min. Move the supernatant into a 250 mL tube to precipitate the AAV1 with high salt. Keep the cell pellets in the 50 mL tubes at 4 °C for AAV1 purification.
   (f) On the sixth day, resuspend the cell pellets in a volume of hypotonic buffer five times the volume of the packed cell volume (about 30 mL), and then add 0.11 volume of 10× restore buffer (about 3.3 mL) and mix by pipetting, and then spin down the nuclei at 2600×*g* for 10 min.
6. Purification of AAV
   (a) Resuspend the nuclei in 1 mL of low salt buffer per 15 cm dishes and transfer it into a 15 mL tube, vortex immediately, and quickly pipette 6 mL of high salt buffer per 15 cm dishes into the tube, and the mixture turns viscous. Sonicate the mixture to break up the genomic DNA until the lysate turns non-sticky (the output at 23%), sonicate at 30 and 10 s interval for about 8 min, and then spin at 2600×*g* for 10 min. The supernatant will be incubated with DNase and RNase at 37 °C for 30 min (0.1 mg per enzyme per 15 cm dish; DNase and RNase stock is 1 mg/mL, so 1 mL each enzyme for ten 15 cm dishes).
   (b) Prepare the discontinuous iodixanol gradients: 9 mL of the 15% layer, 6 mL of the 25% layer, 5 mL of the 40% layer, and 4 mL of the 60% layer.

(c) Sequentially in Beckman OptiSeal ultracentrifuge tubes (*see* **Note 7**).

(d) Marker the interspace between the 60% layer and the 40% layer.

(e) Ultracentrifugation at 350,000×*g* (Beckman VTi 50 rotor, Optima XE-90 ultracentrifuge) at 18 °C for 90 min, and the viral solution should sit between the 40% and 60% layer.

(f) Use a 5 mL syringe attached with a 18G needle to pierce into the centrifuge tube with the needle hold facing the 40% layer.

(g) Slowly draw the viral solution out and put into the 15 mL tube with 10 mL 0.001% F-68/PBS. The mixture will be transferred into a 50 mL filter unit (UFC910008, 100KMWCO) for overnight at 4 °C.

(h) Spin at 4000×*g* in a fixed angle rotor centrifuge for 45 min with the flat surface of the filter membrane perpendicular to the axle of the rotor.

(i) Discard the flow-through and add 10 mL 0.001% F-68/ PBS to the top of the filter, and then centrifuge it again. Repeat until the retention solution becomes clear and the retention volume is around 0.5–1 mL.

7. Quantification of AAV

(a) Set up DNA standards, which are prepared in triplicate in separate 1.5 mL microcentrifuge tubes and later loaded into a 96-well plate. The concentration of the standard stock solution is used to generate the standard curve after qPCR. We also need to prepare a negative control with no DNA and a positive control with a known concentration of virus sample.

(b) Make serial dilutions as follows: prepare a plasmid first, test the concentration, and dilute to 100 ng/μL [$(100 \times 10^{-9}$g/μL$) \div 5181$ bp $\times$ 650dalton/bp (g/mole)] $\times$ $(6.022145 \times 10^{23})$ molecules/mole $= 1.79 \times 10^{10}$ molecules/μL].

(c) To obtain a solution as $2 \times 10^{9}$ molecules/uL: $1.79 \times 10^{10} \div 2 \times 10^{9} = 8.95$ dilution.

(d) Then pipette 10 μL of qPCR master mix plus 10 μL of each standard and diluted sample into each well. Place the plate into the qPCR machine, following the cycling parameters indicated in Table 4.

(e) Calculate the titer (vg/mL) of each virus based on the standard curve (*see* **Note 8**).

**Table 4**
**Cycling parameters for viral titration**

| Step | Denature anneal | Extension |
|---|---|---|
| 1 | 95 °C | 3 min |
| 2 | 95 °C | 5 s |
| 3 | 60 °C | 60 s |
| | Repeat steps 2–3 40× | |
| 4 | 95 °C | 15 s |
| 5 | 60 °C | 30 s |
| 6 | 60 °C | 5 s |
| 7 | 95 °C | 50 s |

### 3.3 In Vivo Genome Editing for Inhibiting Pathological Angiogenesis

1. A mouse model of oxygen-induced retinopathy (OIR) is chosen for the detection of in vivo editing efficiency in inhibiting angiogenesis. In this model, the postnatal day 7 (P7) mice experience 75% oxygen until P12 when the mice come out from the high-oxygen chamber into a normal room air. Due to a relative low level of oxygen, retinal angiogenesis begins to form at P12 and reaches a peak at P17.
2. The dual AAV1s are intravitreally injected into P12 mice under a microsurgical microscope after anesthesia. During injection, the treated eye is punctured at the upper nasal limbus using an insulin injection needle, and then a volume of 1 μL of solution (e.g., AAV1-Cas9 and AAV1-sgRNA, $3 \times 10^{12}$ vg/mL) using blunt glass pipettes with a diameter of approximately 150 μm at the tip or with blunt 32G microliter syringe. The injected solution should be allowed to diffuse over 10 s (*see* **Notes 9** and **10**). After the intravitreal injection, the eye is treated with a triple antibiotic (Neo/Poly/Bac) ointment. After recovery from anesthesia, the mouse is placed in a clean cage and let stay in room air until P17.
3. The injected mice are euthanized at P17, and their retinas are carefully isolated for the following experiments: (1) PCR amplify the desired fragments from their retinal genomic DNA for next-generation sequencing for the analysis of target and off-target effects; (2) Western blot analysis of their retinal lysates for target protein depletion; and (3) whole mount retina staining of their retinas with an endothelium-specific marker (isolectin B4) to analyze pathological and physiological retinal angiogenesis.

## 4 Notes

1. Alternatively, this reaction can be taken in a thermocycler using the following parameters: 37 °C for 30 min, 95 °C for 5 min, and then ramp down to 25 °C at 5 °C/min.
2. Prepare a negative control reaction with digested fragments without annealed oligos.
3. Cell state also affects the transfection efficiency, so cells must not be at full confluency at the time of transfection.
4. DNA must be pure enough via using endotoxin-free plasmid purification kits because the bacterial endotoxin can contaminate cells and reduce transfection efficiency.
5. Mutation in ITR of AAV plasmids can seriously influence transfection, so the ITR plasmids should be verified by sequencing or restriction digestion.
6. During transfection, DNA-PEI complexes must have formed by vortex for 10 s and incubated at RT for 20–30 min.
7. It should leave 12 mL space above the 15% layer for the cellular lysate.
8. AAV titration is also correctly performed; a positive control should be included during AAV titration.
9. It is easy to damage the lens because the spherical lens of the mice is relatively large. Thus, during intravitreal injection the needle towards retrobulbar should be placed at reverse 45 angle instead of vertical injection along limbus.
10. The injected virus might leak out, so during intravitreal injection, the needle must stay for several seconds to prevent the virus from coming out.

## Acknowledgments

This work was supported by the National Natural Science Foundation of China (81900893) to WW, the Science and Technology Plan Project of Hunan Province (2019RS2011) (WW), the National Natural Science Foundation of China (82070989) to HL, and Introduction Plan of High-Level Foreign Experts (G2022026027L) to HL. No funding bodies had any role in the study design, data collection and analysis, decision to publish, or preparation of the manuscript.

## References


1. Nathan DM (1993) Long-term complications of diabetes mellitus. N Engl J Med 328(23): 1676–1685. https://doi.org/10.1056/NEJM199306103282306
2. Nentwich MM, Ulbig MW (2015) Diabetic retinopathy – ocular complications of diabetes mellitus. World J Diabetes 6(3):489–499. https://doi.org/10.4239/wjd.v6.i3.489
3. Kimoto K, Kubota T (2012) Anti-VEGF agents for ocular angiogenesis and vascular permeability. J Ophthalmol 2012:852183. https://doi.org/10.1155/2012/852183
4. Zhao YSR (2018) The role of anti-vascular endothelial growth factor (anti-VEGF) in the management of proliferative diabetic retinopathy. Drugs Context 7:212532
5. Wang D, Zhang F, Gao G (2020) CRISPR-based therapeutic genome editing: strategies and in vivo delivery by AAV vectors. Cell 181(1):136–150. https://doi.org/10.1016/j.cell.2020.03.023
6. Ran FA, Cong L, Yan WX, Scott DA, Gootenberg JS, Kriz AJ, Zetsche B, Shalem O, Wu X, Makarova KS, Koonin EV, Sharp PA, Zhang F (2015) In vivo genome editing using staphylococcus aureus Cas9. Nature 520(7546): 186–191. https://doi.org/10.1038/nature14299
7. Li F, Wing K, Wang JH, Luu CD, Bender JA, Chen J, Wang Q, Lu Q, Nguyen Tran MT, Young KM, Wong RCB, Pebay A, Cook AL, Hung SSC, Liu GS, Hewitt AW (2020) Comparison of CRISPR/Cas endonucleases for in vivo retinal gene editing. Front Cell Neurosci 14:570917. https://doi.org/10.3389/fncel.2020.570917
8. Huang X, Zhou G, Wu W, Duan Y, Ma G, Song J, Xiao R, Vandenberghe L, Zhang F, D'Amore PA, Lei H (2017) Genome editing abrogates angiogenesis in vivo. Nat Commun 8(1):112. https://doi.org/10.1038/s41467-017-00140-3
9. Huang X, Zhou G, Wu W, Ma G, D'Amore PA, Mukai S, Lei H (2017) Editing VEGFR2 blocks VEGF-induced activation of Akt and tube formation. Invest Ophthalmol Vis Sci 58(2):1228–1236. https://doi.org/10.1167/iovs.16-20537
10. Wu W, Duan Y, Ma G, Zhou G, Windhol C, D'Amore PA, Lei H (2017) AAV-CRISPR/Cas9-mediated depletion of VEGFR2 blocks angiogenesis in vitro. Invest Ophthalmol Vis Sci 58(14):6082–6090. https://doi.org/10.1167/iovs.17-21902

# INDEX




Guei-Sheung Liu and Jiang-Hui Wang (eds.), *Diabetic Retinopathy: Methods and Protocols*, Methods in Molecular Biology, vol. 2678, https://doi.org/10.1007/978-1-0716-3255-0,

MIX
Papier aus verantwortungsvollen Quellen
Paper from responsible sources
FSC® C105338

If you have any concerns about our products,
you can contact us on
**ProductSafety@springernature.com**

In case Publisher is established outside the EU,
the EU authorized representative is:
**Springer Nature Customer Service Center GmbH**
**Europaplatz 3, 69115 Heidelberg, Germany**

Printed by Libri Plureos GmbH
in Hamburg, Germany